BIOLOGY 185
Laboratory Manual & Student Study Guide

Bauman, Robert, *Microbiology With Diseases by Taxonomy*,
3rd Edition, Pearson Benjamin Cummings, San Francisco, CA, 2007.

Robert Bauman (left)
Robert Brader (right)

Third Edition
Robert E. Brader
Microbiology Instructor
Andrew B. Brader
Microbiology Instructor

Lancaster General College
of
Nursing & Health Sciences

Kendall Hunt
publishing company

www.kendallhunt.com
Send all inquiries to:
4050 Westmark Drive
Dubuque, IA 52004-1840

Printed in the United States of America
10 9 8 7 6 5 4 3

"Ring around the rosie, a pocket full of posies. Ashes, ashes, we all fall down"

A nursery rhyme from about 1347, derived from the Black Plaque, which killed over 25 million people in the 14th century. The "ring around the rosie" refers to the round red rash that is the first symptom of the disease. The practice of carrying flowers and placing them around the infected person for protection is described in the phrase, "a pocket full of posies." "Ashes" is an imitation of the sneezing sounds made by the infected person. Finally, "we all fall down" describes the many dead resulting from the disease (Kate Greenaway 1846-1901).[1]

This is another example of an "urban legend" described by Jan Harold Brunvand, a retired professor from the University of Utah, whose 2-volume set, Urban Legends and his current reference set; The Truth Never Stands in the Way of a Good Story provide insightful reading.[2] "Ring around the Rosie" is just a nursery rhyme, appearing originally in Kate Greenaway's Mother Goose or "The Old Nursery Rhymes" in 1881. Probably the reference is to be found in the religious ban on dancing among many Protestants in the 19th century, in Britain as well as in North America. Adolescents found a way around this dancing ban by what was called in the United States the play party. Play parties consisted of ring games, which were hugely popular and younger children also partook of the act. Some modern nursery games, particularly those involving rings of children, were derived from these play party games. *Little Sally Saucer* was one of them, and *Ring Around the Rosie* seems to be another.[3]

Robert E. Brader & Andrew B. Brader

INITIAL PRINTING
June 2003

SECOND PRINTING
August 2003

THIRD PRINTING
May 2004

FOURTH PRINTING
December 2004

FIFTH PRINTING
December 2006

SIXTH PRINTING

[1]Greenaway, K. (1881). *Mother Goose or the Old Nursery Rhymes*. London, UK: George Rutledge & Sons.

[2]Brunvand, J.H. (1994). <u>The Truth Never Stands in the Way of a Good</u>, *The Big Book of Urban Legends*. New York, NY: Paradox Press.

[3]Varasdi, A. (1989). *Myth Information*. New York, NY: Ballantine Books.

PREFACE

Victor Hugo so aptly stated, "Where the telescope ends, the microscope begins, which of the two has the grander view?"

Welcome to the world of **Microbiology**. It is our hope that you, as a beginning professional in the field of health sciences, acquire an understanding and appreciation for the complex interaction between man and microbes.

The intent of this study guide is to provide you with a supplement for the lecture portion of the course, a systematic approach to case studies in Microbiology, and a comprehensive list of all the laboratory assignments required throughout the semester. It is not meant to be a substitute for attending lectures and labs, since additional material will be presented in both. A worthwhile supplement for this course is the *Microbiology Coloring Book*, which can be purchased at any bookstore. This activity, as simple as it may seem, has proven to be a highly-effective method for learning reinforcement and retention enhancement. Actively coloring, rather than only passively reading a textbook, provides a more intense focus on Microbiology topics.[1] A valuable website www.microbiologyplace.com. is provided the student with the purchase of the textbook. This adjunct to the text enlightens as well as reinforces concepts in the fascinating field of microbes.

Robert E. Brader Brader
BA Millersville University University
MA Temple University

Andrew B.

BS Virginia Tech

MS New York University

---

[1] Alcamo, E., & Elson, M. (1996). *The Microbiology Coloring Book*. San Francisco, CA: Benjamin Cummings.

MICROBIOLOGY LABORATORY

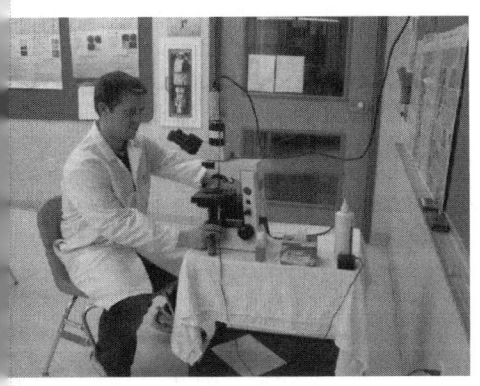

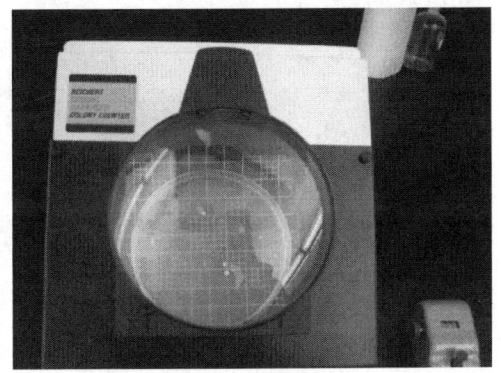
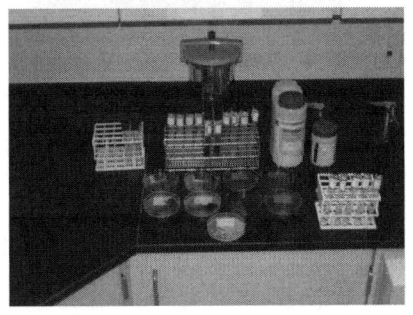
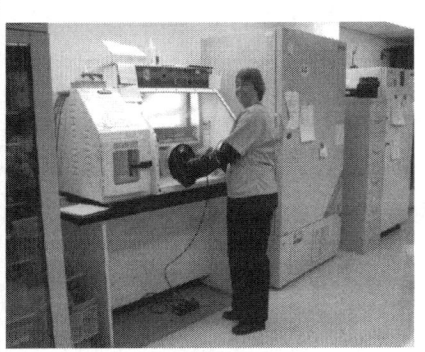
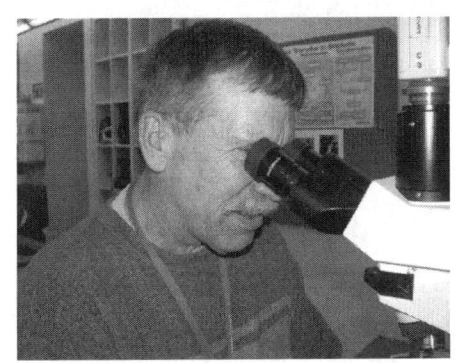
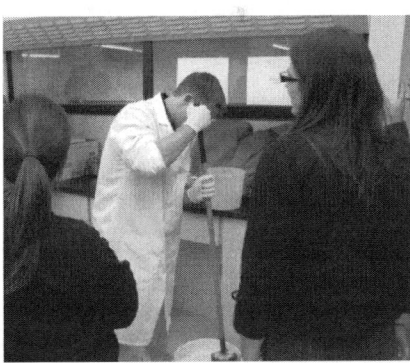

TABLE OF CONTENTS

LABORATORY EXERCISES

LABORATORY SUPPLEMENTAL EXERCISES

MICROBIOLOGY FIELD TRIPS

LECTURE OUTLINE NOTES

LABORATORY RULES

1. Eating or drinking of liquids is not permitted in the Microbiology Laboratory. It is also not advisable to have any food or drinks in the Laboratory, even if stored in backpacks. The cultures we will be using are considered non-pathogenic; however, any organism has the capability of being an opportunistic pathogen and thus can cause problems under certain conditions.

2. Always use the microscope assigned to you unless otherwise directed. Note: Brightfield microscopes are expensive, so please use care in the transport to and from the bench top.

3. Clean and store your microscope in the fashion described in class.

4. Limit personal belongings on the laboratory bench. A clean bench is a safe bench. Upon entering the lab, place coats, books, and other paraphernalia in designated locations – NEVER on bench tops.

5. Wash hands frequently, using detergents provided. Eighty percent of all diseases are transmitted via contact from hands to mouth/nose or hands to eyes.

6. Report any accidents to your instructor immediately. Any accident, no matter how insignificant, can be a problem.

7. Report the breakage of any laboratory items to your instructor immediately and always dispose of any sharp material in the "sharps container."

8. All slides and broken glassware that is to be disposed of NEEDS to be discarded in the BROKEN GLASS boxes at the front of the lab. Do NOT throw paper towels in this box.

9. Use caution when using the Bunsen or Meeker burners. Note: NEVER leave a flame unattended.

10. Wipe down laboratory bench areas before and after laboratory sessions with disinfectant/sanitizer provided.

11. Safety glasses should be used when working in the laboratory, especially when sterilizing loops or needles with the Bunsen burner.

12. Begin cleanup of your laboratory work area 5-10 minutes before the end of the session. Dispose of all cultures and materials as designated by your instructor.

13. Store materials as directed by your instructor. Never remove bacterial cultures from the laboratory. STRICTLY PROHIBITED

14. Do not store wrappers, used lens paper, and miscellaneous trash in drawers.

15. __IN THE LABORATORY NEATNESS DOES COUNT.__

Laboratory #1
The Microscope and You

The eye of a human being is a microscope, which makes the world seem bigger than it really is.

Kahlil Gibran

Introduction: Microbiology, the branch of science that has so vastly extended and expanded our knowledge of the living world, owes its existence to Antoni van Leeuwenhoek. In 1673, with the aid of a crude microscope consisting of a biconcave lens enclosed in two metal plates, Leeuwenhoek introduced the world to the existence of microbial forms of life. Over the years, microscopes have evolved from the simple, single-lens of Leeuwenhoek, with a magnification of 300X, to the present day electron microscopes, capable of magnifications greater than 1,000,000X. Microscopes are designed as either light microscopes or electron microscopes, the former using visible light or ultraviolet rays to illuminate specimens. Electron microscopes use electron beams instead of light rays to observe submicroscopic particles. We will be using the brightfield compound microscope.

Anatomy of the Microscope:
CX31 Olympus Biological Microscope

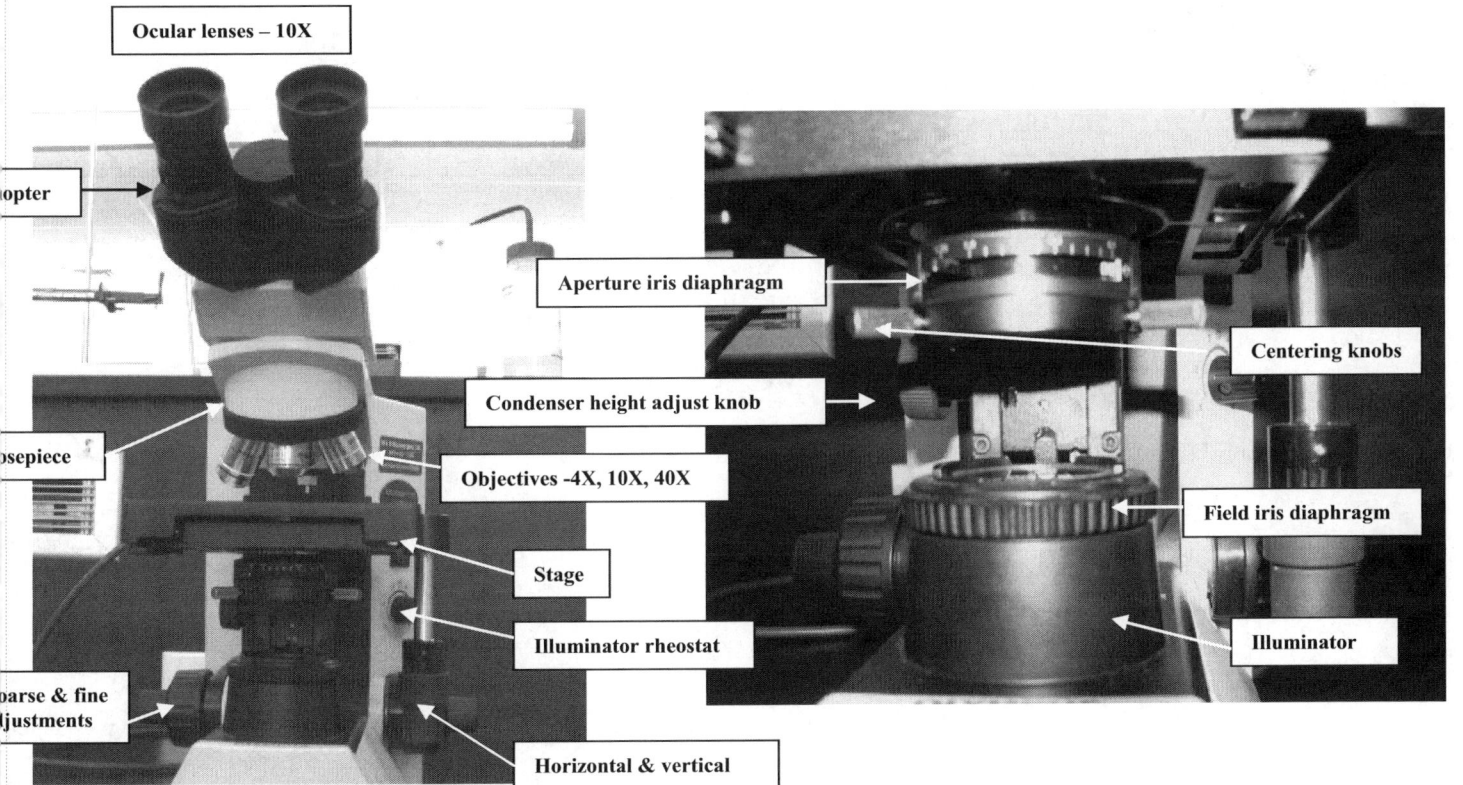

Theoretical Principles of Microscopy:

Students should understand magnification, resolution, numerical aperture, illumination, and focusing for proper use of the microscope. **Magnification** is the enlargement of a specimen and is a function of the two lenses in a compound microscope (ocular lens and objective lens). The objective lens enlarges the specimen and produces a real image projected inside the body tube or barrel. The ocular lens then further enlarges this image to produce a final image (or **virtual image**) that we view. **Resolution or resolving power** is the ability of a lens system to display two adjacent objects as discrete entities. If the two objects appear as one then the lens system has lost its resolution. Increased magnification will not rectify the situation. In fact, it will blur the object even more. **Numerical aperture** is defined as a function of the diameter of the objective lens in relation to its focal length. **Illumination** is the light necessary to view the specimen. **Focusing** is the proper adjustment of the lens system to the specimen in order to obtain a clear distinct image.

Rheostate – dimmer (between 3 + 4)

Key Functions of the Microscope:

1. **Illumination**-daylight is an uncontrolled variable; therefore, we use a tungsten lamp, which has a specific wavelength. There are two controls for the intensity of the light, a rheostat and a diaphragm.

2. **Magnification**-ocular power times the objective power *(lens) = total*

3. **Resolution**-$R = 0.61 \times$ Wavelength of light/$NA_{objective}$ *= Distance*

4. **Maximum magnification**-Ocular lens (10X) X Oil Immersion (100X) = 1000

5. **Maximum resolving power**-$0.61 \times 410nm/1.25 = 200nm$ or 0.20micrometers

6. **Oil immersion lens**-as light passes from one medium (glass) to another medium (air) the light rays bend (refractive index). Since high illumination is needed with high magnification, some of the light rays will not enter the objective lens due to this bending, therefore a drop of oil between the slide and the objective lens prevents this loss of light. The oil has the same refractive index as glass and thus provides a continuous path for the light.

Distance between 2 pts that can be distinguished as 3 separate entities.

$$\frac{Wavelength \times 0.5ul}{N.A.}$$

NA = numerical Aperture

Empty mag – magnification w/out resolution

Parfocalization - + lens in focus, all in focus w/ minor adjustment

Refractive Index-because the refractive indices of the glass, microscope slide, and immersion oil are the same, the oil prevents the light rays from refracting

light source diaphram- open full

condenser - 2 screws on it focuses light into specimen

Condenser diaphram ~ 1.25

Light Spectrum

infared

electron
x Ray
Gamma Ray

large Radio wave

ROYGBIV

UV Rays

Small

* If wavelength smaller better resolution

✱ Resolution= Smallest distance between 2 pts. that we can still distingish as a entitia

Usage of the Microscope:

I. **Proper illumination**

Quick Set Up
1. Set Illuminator Rheostat to three quarter Open
2. Move Condenser to Top of Stage
3. Set Aperture Diaphragm to 1.0
4. Adjust Diopter Adjustment for Left Eye
5. Open Field Iris Diaphragm so Light Fills View, Then Open Slightly More

II. **Operation of the microscope:**

all lens in focus

Parfocalization-when one lens is in focus, the other lenses will also have the same focal length and can be rotated into position without further *major adjustments*. In practice, however, usually a one half turn of the fine adjustment knob in either direction is necessary for sharp focus.

Begin focusing the microscope with either the scanning lens or the low power lens, both considered safe lenses, by lowering the body tube to its lowest position with the coarse adjustment knob. Using the fine focus knob, sharply focus the specimen. Move up to the next higher magnification and only focus with the fine focus knob. Finally rotate objectives to allow an opening in front of the microscope and place a drop of immersion oil on the slide. Rotate the oil immersion lens into the oil and sharply focus with the fine focus knob only.

Red + yellow lens considered Safe
wipe oil off lens when done
once under oil DO NOT go back to Blue lens
only use fine adjustments w/ Blue + white

3 ways to adjust light

1.) Rheostat - 3+4

2.) Condenser diaphram - All the way to left

3.) light source diaphram - Full open

– moves stage

Calibration of the Ocular Micrometer using a Stage Micrometer:

1.   *Observe the stage micrometer using the 4X objective and center in the field – **SEE BELOW***

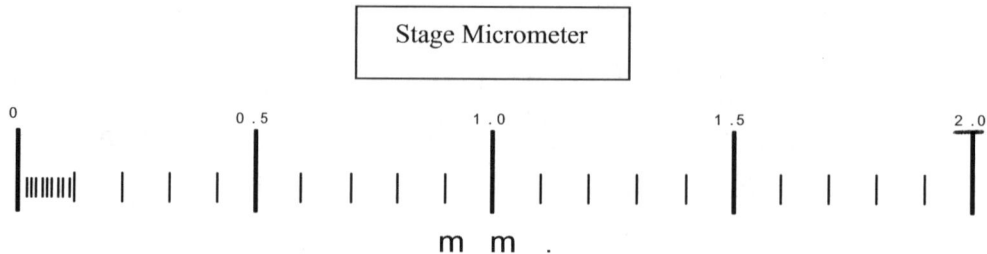

Large unmarked lines = 0.1mm and small unmarked lines between 0 and 0.1mm = 0.01mm.   1mm = 1000micrometers or 1micrometer = 0.001mm.

2.   *Next observe with the 10X objective and center the Vernier scale in the field. Overlay the eyepiece micrometer on top of the stage micrometer, with 0 on 0 and you see that every 10 lines on the eyepiece micrometer = 0.1mm on the stage micrometer (100micrometers) or every line on the eyepiece micrometer = 0.01mm on the stage micrometer (10micrometers).*

3.   *Observe with the 40X objective and you see that 40 lines on the eyepiece micrometer = 0.1mm (100micrometers), so 4 lines on the eyepiece micrometer= 0.01mm (10micrometers). Thus 1 line on the eyepiece micrometer = 0.01mm/4 or 0.0025 mm (2.5micrometers).*

4.   *Observe with the 100X objective (oil immersion) and you see that 100 lines on the eyepiece micrometer =0.1mm (100micrometers), or 10 lines = 0.01mm (10micrometers), thus 1 line = 0.001mm (1micrometer).*

In this class if you remember anything about calibration, remember that when using the 100X objective (oil immersion), which will be most of the time, every hash mark on the ocular micrometer is equal to 1micrometer.

Microscope Storage:

1.   Move the scanning lens to position over the stage.
2.   Wipe the oil from the oil immersion lens with lens paper only.
3.   Wrap the cord around the base and cover the microscope.

Microscope Considerations:

1.   Do not let oil get on any of the other lenses, since they are not sealed. If oil does get on the lens, wipe off immediately with lens paper, but also tell the instructor.
2.   Do not use any cleaning solvents on the objective lenses. If the lenses need to be cleaned see the instructor.
3.   Transport the microscope as shown and keep the microscope back from the front of the lab bench when using it.

1 mm = 1000 micron

1mm = Bacteria

Objective Lens Specifications

Condenser
Diaphram

Name	Color Code	Power	Total Magnification	Numerical Aperture	Aperture Diaphragm Setting	Working Distance
SCANNING LENS	Red	4x	40	.1	1.0 250μm	25mm micro (30mm)-WD 40/1000
LOW DRY POWER	Yellow	10x	100	.25	1.0 10μm	10micron (6-10mm)-WD 100 ticks/1000 micron
HIGH DRY POWER	Blue	40x	400	.65	1.0 2.5mm	2.5 micro (.6mm)-WD 40/100
OIL IMMERSION	Black	100x	1000	1.25	1.0 1μm	1 micron (.18 mm)-WD

Laboratory #2
Microscopy

The most important discoveries of the laws, methods, and progress of nature have nearly always sprung from the examination of the smallest objects she contains.

Jean- Baptiste LaMarck

Beginning students frequently become impatient with the microscope and forgo this opportunity to practice and develop their observation skills. Observation requires patience and curiosity. Make careful sketches and pay special attention to the following:

* **Size Relationship:** How big are bacteria in relationship to blood cells, protozoa, and algae?
* **Spatial Relationship:** Where is one bacterium in relation to the others? Are they together, separate, or in chains?
* **Behavior:** Are individual cells moving, or are they all flowing in the medium (*Brownian motion*)?
* **Sequence of Events:** Were cells active when you first observed them?

I. Human Blood Smear (**Wright's Stain**)
   A. Prepared slide
   B. Place on stage
   C. Progress from Scanning Lens to High Dry Power and finally to Oil Immersion
   D. Measure erythrocytes, leukocytes, platelets
   E. Wipe off oil from Oil Immersion lens and slide

II. Microbial Classification - Prepared Slides – **Animal-like Group - Protozoa**

EUKARYOTES/PROTOZOANS

A.	**Sarcodina**	(Pseudopodia motility)
		Amoeba proteus
B.	**Mastigophora**	(Flagella motility)
		Trypanosoma lewisi (sleeping sickness)
C.	**Ciliophora**	(Cilia motility)
		Paramecium caudatum
		Blepharisma
		Stentor
D.	**Sporozoa**	(Nonmotile)
		Plasmodium vivax (malaria) - **Information only**

III. Microbial Classification - Prepared Slides - **Plant like Group**

EUKARYOTES

A.	**Euglenoid**	
		Euglena
B.	**Algae-single cell**	
		No representative specimen
C.	**Algae-filamentous**	
		Spirogyra vegetative

IV.   Microbial Classification - Prepared Slides - **Bacteria**

PROKARYOTES

    A.   **Bacillus** - rod shaped organisms
                Bacillus subtilis
    B.   **Cocci** - ball shaped organisms
                Staphylococcus aureus
    C.   **Cyanobacteria** (used to be blue-green algae)
                Anabaena (filamentous in chains)

V.   Microbial Size

    A.   Measurement-Metric System

1 millimeter (mm) = 1,000 micrometers (μm)
1 millimeter (mm) = 1,000,000 nanometers (nm)
1 micrometer (μm) = 1,000 nanometers (nm)

    B.   Approximate size of microbes

Bacteria = 1-3 micrometers (μm)
Yeasts = 2-3 micrometers (μm)
Paramecium = 200 micrometers (μm) or 0.2 millimeters (mm)
Viruses = 10-300 nanometers (nm) or 0.01–0.3 micrometers (μm)

    C.   Measure using ocular micrometer (*Previous Lab Calibration*)

LABORATORY #2
Microbial Observation

Name_____

Date_____

High Dry Objective: *1 division = 2.5 μm* **Oil Immersion:** *1 division = 1.0 μm* *(distance of graduations on ocular micrometer)*

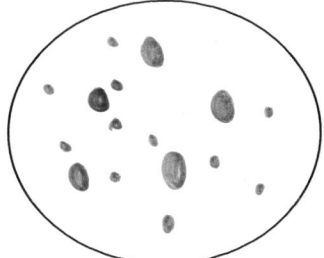

Blood Smear:
Size: Erythrocytes:_____
     Leukocytes:_____
     Platelets:_____
Power 40x

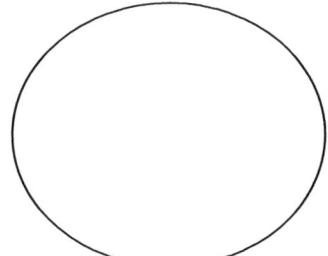

Protozoa:
Cyanobacteria
 Size_____
Power_____

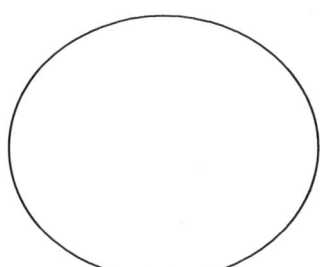

Algae:
Spirogyra
Size_____
Power_____

Protozoa:
Amoeba proteus
Size 100 Mm
Power 10x

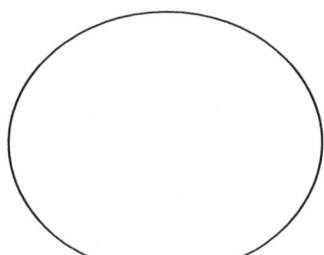

Protozoa:
Paramecium caudatum
Size_____
Power_____

Protozoa:
Trypanosoma lewisii
Size 1500 Mm  15mm
Power 100x

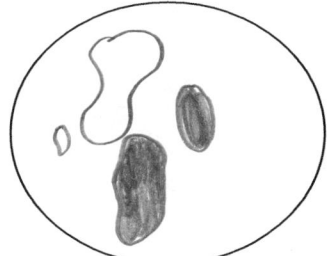

Bacteria:
Staphylococcus aureus
Size 1200 Mm   120mm
Note size relative to erythrocytes
Power 100x

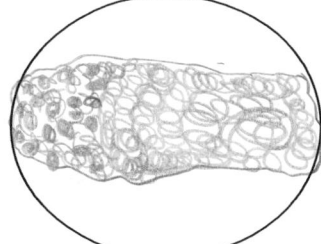

Bacteria:
Bacillus subtilis
Size 100,000 Mm/100mm
Note size relative to erythrocytes
Power 100x

motility

* Living Organisms — How much light is being absorbed
  - 70% Water : Refractive index is similar to environment (Hard to see)

* Methods (live)
  - Wet mount - easiest way to see live organisms.

  - Hanging drop * NO oil why?

  W.D = 0.18mm
  (Break coverslip)

- movement (flagella)
  - Eukaryotes - Flagella undulate. (Flipper)
    - Smooth motion   - Cilia
                       - Psedopods
      - no nucleus   - No movement
  -* (Bacteria) Prokaryotes - Flagella rotate (propeller) Boat
      - Erratic motion   Runs + tumbles

-* Brownian motion - False movement, has to do w/ how small it is. (like vibrating) water molecules hitting bacteria.

Laboratory #3
Microscopy - Motile Cultures

Everything is in motion. Everything flows. Everything is vibrating.

William Hazlitt

Objective: To differentiate the differences between eukaryotic and prokaryotic motility. To differentiate **Brownian motion**[1] from true motility.

Eukaryotic motility

Sarcodina: Pseudopodia motility
Example*: Amoeba proteus*

Mastigophora: Flagella motility
Example: *Euglena*

Ciliophora: Cilia motility
Example: *Paramecium , Blepharisma, Stentor*

Sporozoa: Non- motile
Example: *Plasmodium spp.* (malaria) demonstration

Prokaryotic motility--always associated with flagella

Monotrichous   Amphitrichous   Lophotrichous

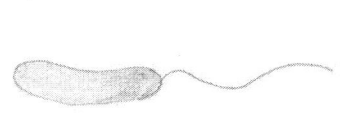   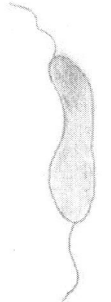   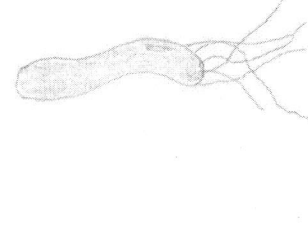

Peritrichous   Axial filament

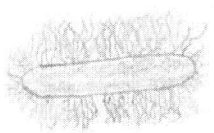   

[1]Brownian Motion: Observed by Robert Brown in 1827, while looking at dead 100- year- old pollen. Random oscillating motion caused by molecules of solvent bombarding the ultra -small objects.

Procedure:

Wet Mount: use for live preparation of eukaryotes[1]

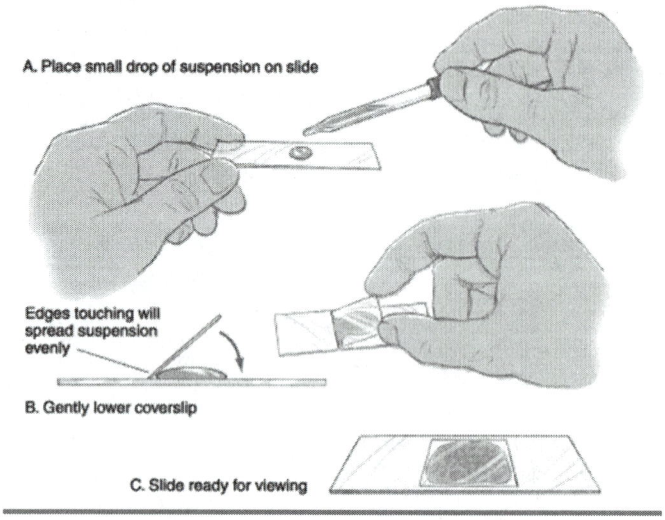

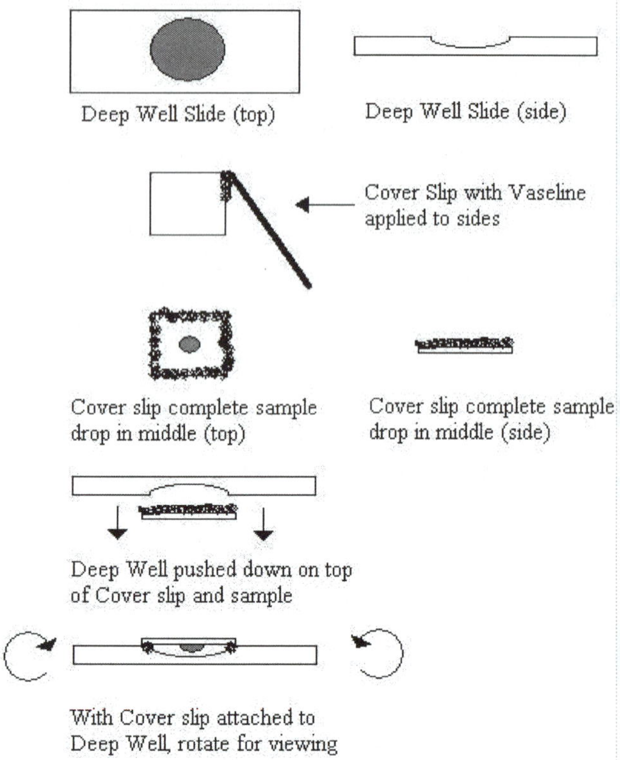

A. Place small drop of suspension on slide
Edges touching will spread suspension evenly
B. Gently lower coverslip
C. Slide ready for viewing

Hanging Drop Preparation: use for live preparation of prokaryotes

Deep Well Slide (top)
Deep Well Slide (side)
Cover Slip with Vaseline applied to sides
Cover slip complete sample drop in middle (top)
Cover slip complete sample drop in middle (side)
Deep Well pushed down on top of Cover slip and sample
With Cover slip attached to Deep Well, rotate for viewing

Semi-solid agar deeps: use for live preparations of prokaryotes to determine motility

Non-motile | Motile
1st & 2nd tubes | 3rd tube

[1]© Kendall Hunt Publishing Company

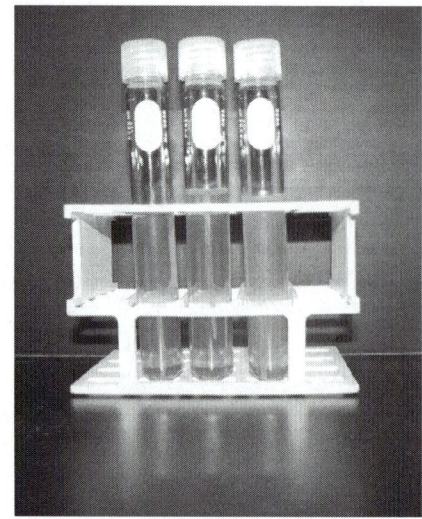

use red + yellow

Laboratory #3
Microbial Observation

Name: _____

Date: 9/2/14

( )

Protozoa (wet mount)
Describe briefly:

( )

Protozoa (wet mount)
Describe briefly:

( )

Protozoa (wet mount)
Describe briefly:

( )

Bacteria (hanging drop)
Describe briefly:

Bacteria in semi-solid agar
Describe briefly:

Organisms	Shape	True Motility or Brownian
Staphylococcus aureus		
Escherichia coli		

1. Why are microorganisms hard to see in wet preparations?

2. How do you determine motility in semi-solid agar? Why use such a low concentration of agar?

Laboratory #4
Simple Stain Technique

Chance favors the prepared mind.

Louis Pasteur

I. **Staining Techniques to View Bacteria**
   A. **Simple stain-**(basic or acidic/negative stain)
   B. **Differential stain-**(gram stain, acid fast stain)
   C. **Structural stain-**(spore stain, flagella stain, capsule stain)

II. **Simple stain-Lab #4**
   A. Bacteria are difficult to distinguish from their surroundings-approximately the same refractive index as surroundings-appear almost colorless.
   B. Staining provides a method to visualize either the whole organism or specific parts of the organism, by changing the refractive index and thus enhancing the contrast between the organism and its surroundings - *observe greater detail and resolution than wet mount.*
   C. Stains are synthetic aniline (coal tar derivatives) dyes derived from benzene. Most dyes are salts, composed of charged colored ions, called **chromophores.**
      1. *Basic stain* - chromophore is the positive ion-cation (most of dyes used in microbiology).
         (SALT) dissociates into a (CATION) and (ANION)
         Methylene blue chloride----------Methylene blue$^+$ + Cl$^-$
                                            (Chromophore)   (Anion)
         Acidic stain - chromophore is the negative ion-anion
   D. Direct stain (simple stain)
      1. Most bacteria have a slightly negative charge-due to cell wall and cytoplasmic constituents.
      2. Basic stains are attracted to the negative charge and bind.
         a. *Methylene blue*
         b. *Crystal violet*
         c. *Safranin*
         d. *Carbol fuchsin*
         e. *Malachite green*
   E. Negative stain (simple stain)
      1. Acidic stains are repelled by the negative charge of a cell but are attracted to the slightly positive charge of the surrounding environment.
      2. Background stained but bacteria remain unstained.
         a. *Nigrosin*
         b. *Eosin (sodium eosinate)*

III. **Staining Procedure**
   A. Stain preparation (on bench)
      1. Clean and label slides--*note: slides are generally pre-cleaned*
         a. Clean and dry slides - handle clean slides by edge or end.
         b. Use marker to make dime-sized circles on each slide, on the bottom of the slide so they will not wash off--label each as to the culture used.
      2. Deposit inoculum - *Smear*

a.   Solid media cultures--place 1 loop of water in center of circle - use sterile loop to scrape off a tiny amount of culture - emulsify thoroughly in water (*smear should look like diluted skim milk*) and spread throughout circle.

b.   Broth media cultures--*do not use water*, because bacteria are already suspended in water - re-suspend bacteria in broth by vortexing--use sterile loop to deposit 1 loop of culture in the circle and spread throughout circle.

3.   Dry slides (air dry)--do not blow on slides as this will move bacterial suspensions.

4.   Heat fix slides (3X through flame).

B.   Staining (on tray)

1.   Apply stain to cool slide (60") be consistent with time.

2.   Wash the slides--beaker of water or wash bottle-- allow to run down tilted slide.

3.   Gently blot dry with bibulous paper or paper towel – *do not wipe*.

IV.   Observation - Microscope

A.   Work to oil--*No Coverslips*

B.   Sketch

C.   Label

1.   Magnification

2.   Specimen and dye

3.   Shape

V.   Negative Simple Stain

Aseptically add a loop of culture to a drop of negative stain, mix, and follow the procedure below, avoiding over inoculating with organisms:

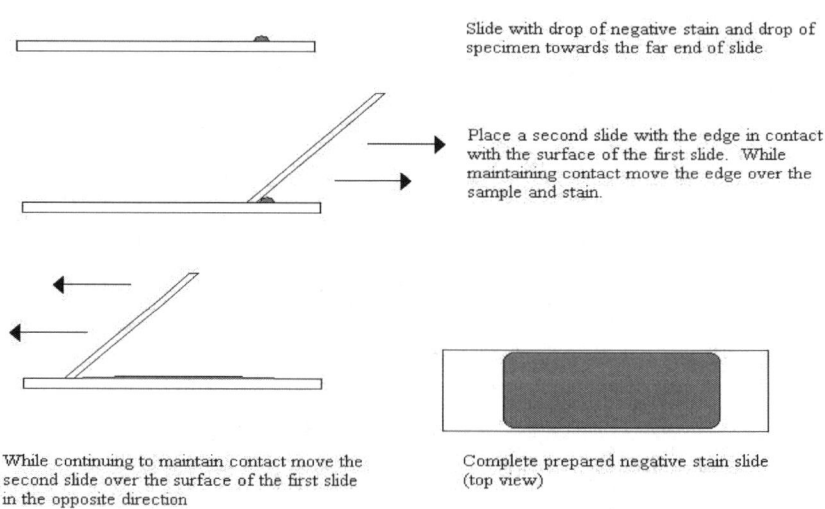

Slide with drop of negative stain and drop of specimen towards the far end of slide

Place a second slide with the edge in contact with the surface of the first slide. While maintaining contact move the edge over the sample and stain.

While continuing to maintain contact move the second slide over the surface of the first slide in the opposite direction

Complete prepared negative stain slide (top view)

Allow to air dry and observe – *Do not heat fix*

Laboratory #4
Microbial Observation

Name: _____

Date: _____

Smear from Plate
Organism:_____

Smear from Broth
Organism:_____

Negative Stain
Organism:_____

Demonstration of 3 morphologies: *(1) **cocci** – Staphylococcus aureus **(2) bacilli** – Escherichia coli **(3)*** ***spirals** – Rhodospirillum rubrum*

1.   Of what value is a simple stain?

2.   What is the purpose of fixing the smear?
        a.
        b.
        c.

3.   What is the value of a negative stain?

Laboratory #5
Differential Stain Technique

Experience is the father of wisdom, and memory the mother.

Thomas Fuller

I.   Staining techniques - **Differential Stain**

    A. The **acid-fast stain** (for information only)--differentiates the *Mycobacteria* from all other organisms because acid-fast organisms have dense waxy peptidoglycan-mycolic acid cell walls.

        1.   Tuberculosis, leprosy, and possibly crohn's are a result of *Mycobacterium spp* infections.

        2.   Steps

            a.   Initial stain with carbol fuchsin

            b.   Wash with HCl solution (differential part)

            c.   Counterstain with methylene blue

            d.   Red in oil immersion field indicates acid-fast organisms

        3.   Two methods

            a.   Ziehl-Neelsen - heat

            b.   Kinyoun - cold

II.   Gram Stain

    A. Developed by Christian Gram in 1884

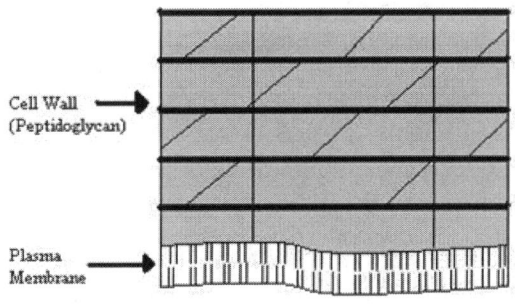

Gram Positive Cell Wall

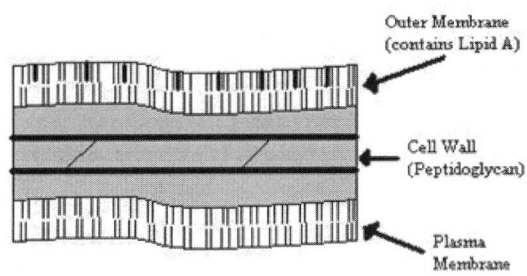

Gram Negative Cell Wall

Bacteria stain differently because of chemical and physical differences in their cell walls. Crystal violet (basic stain) is picked up by the cell's negative attraction. Iodine then reacts with the dye in the cytoplasm to form a crystal violet-iodine (CV-I) complex. In gram-negative cells, the decolorizing agent dissolves the outer LPS layer and the CV-I complex washes out through the thin layer of peptidoglycan. The CV-I complex cannot wash out through the thick layer of peptidoglycan found in gram-positive cells. Gram stain is most consistent when performed on young cultures (<24hours old).

    B. Most important stain in Microbiology

        1. Wide application - initial ID

        2.   Indicates pathogenicity

        3.   Indicates antibiotic regimen

    C. Two methods

        1.   Huckers - crystal violet - USA

        2.   Jensens - methylene blue - European labs

III. Staining Procedure - Huckers'
  A. Stain preparation (on bench)
    1. Clean and label slides - flip over
    2. Deposit inoculum on slide
    3. Dry slides (air dry)
    4. Fix slides (3X through flame)
  B. Staining (on tray)--be consistent with times
    1. Initial stain with crystal violet - 60"
    2. Wash the slides with water - d*o not blot dry*
    3. Mordant - iodine - 60"
    4. Wash the slides with water - *do not blot dry*
    5. Decolorize – alcohol - approximately 5" (differential part)
    6. Wash the slides with water - *do not blot dry*
    7. Counter stain with gram safranin - 60"
    8. Wash the slides with water
    9. Now blot dry with bibulous paper or paper towel *do not wipe*
IV. Observation - Microscope
  A. Work to oil - *no coverslips*
  B. Sketch
  C. Label
    1. Magnification
    2. Specimen and dye
    3. Shape
    4. Gram stain result

Gram Stain Reaction - Viewing at the Different Steps

TIME	GRAM + ORGANISMS (Color Reaction)	GRAM – ORGANISMS (Color Reaction)
Initial Stain		
Mordant-Iodine		
After Decolorizer		
After Counterstain		

Laboratory #5
Microbial Observation

Name_____

Date_____

○   ○   ○

Staphylococcus aureus   *Escherichia coli*   *Bacillus spp.*

Morphology_____   _____   _____

Gram Reaction_____   _____   _____

Which organism is the largest?_____ Smallest?_____

1.   Why will old gram-positive cells stain gram negative?

2.   Suppose you gram stained a sample from a pure culture of bacteria and observed a field of both red and purple cocci. Adjacent cells were not always the same color. What do you conclude?

3.   Since you cannot identify bacteria from a gram stain, why might a physician perform a gram stain on a sample before prescribing an antibiotic?

Laboratory #6
Media & Cultures - Aseptic Technique
Beginning Fermentation

Life would not long remain possible in the absence of microbes.   **Louis Pasteur**

Introduction: In the laboratory, bacteria must be cultured in order to facilitate identification and to examine their growth and metabolism. Bacteria are inoculated, or introduced, into various forms of culture media in order to keep them alive and to study their growth. Inoculations must be done without introducing unwanted microbes, or contaminants, into the media. **Aseptic technique** is used in microbiology to exclude contaminants. All culture media are sterilized prior to use. **Broth cultures** provide large numbers of bacteria in small spaces and are easily transported. **Agar slants** are test tubes containing solid culture media that were left at an angle while the agar solidified. Agar slants provide a solid growth surface, with a large surface area for diffusion of oxygen and a deep butt containing reserves of nutrients. **Agar deeps** are used to grow bacteria that prefer less oxygen than is present on the surface of agar; also **semisolid deeps** (0.5% vs. 1.5%) are used for motility studies. **Agar culture dishes or petri dishes** provide a large flat surface area for growing and observing microorganisms. Transfers of organisms are performed using sterile **inoculating loops** or **inoculating needles** made of nichrome or platinum wire. Transfers can also be made with sterile cotton swabs, pipettes, syringes or glass rods.

I. Microbiological Media
    A. Demonstration - agar and broth
        1. Media in tubes
            a. Broth
            b. Agar
                (1) Slants
                (2) Deeps
        2. Media in culture or petri plates (100 X 15mm plates)
II. Microbiological Media Properties
    A. Agar temperatures
        1. Dissolves and liquefies at 100° C
        2. Solidifies at temperatures below 40° C
    B. Agar concentration
        1. Solid (1.0%- 2.0%)
        2. Semi-Solid (0.05%-0.50%)
    C. Agar is non-digestible and thus remains solidified
    D. Gelatin is digestible by many organisms (12-15%)
        1. Gelatin also liquefies at 35°C
III. Media demonstrations
    A. Patriotic plate = mixed cultures of 3 different organisms and streaked for isolation
                Escherichia coli   (Tan/white)
                Serratia marcescens   (Red)
                Chromobacterium violaceum (Blue)
    B. Pure culture plates streaked for isolation of individual colonies
                Escherichia coli
                Serratia marcescens
                Chromobacterium violaceum

     C.   Inoculated broths

          D.   Selective/differential agar plates

               1.   MacConkey agar - *Escherichia coli*

               2.   Mannitol salt agar - *Staphylococcus aureus*

          Each group will write a very brief description of the media demonstrated.

IV.   Microbiological Culture Techniques - subculturing bacteria

          A.   Aseptic technique (Robert Koch)-demonstration

          B.   Quadrant streak of mixed patriotic plate-sequential dilution

Each student will perform a quadrant streak of the mixed broth culture to produce a patriotic plate visualizing three distinct colonies.

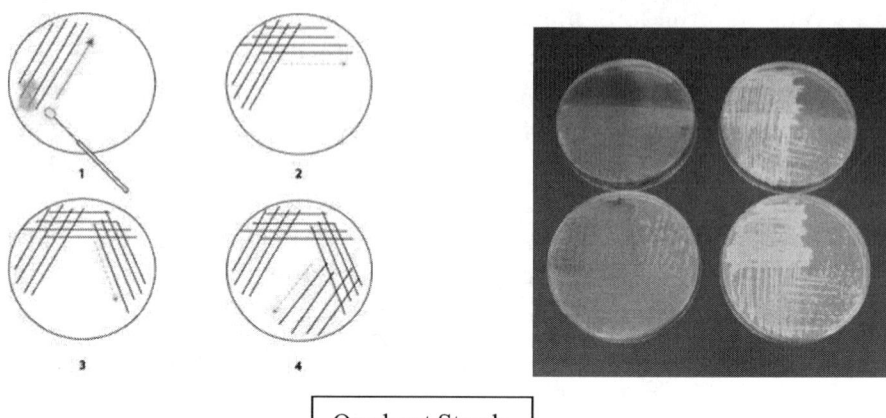

Quadrant Streak

          C.   Heat source

               1.   Bunsen burner or bacticinerator

V.   Fermentation - *WE WILL DEMONSTRATE TWO OR MORE OF THESE PROCESSES IN LAB*

          Metabolic process by which an organic molecule acts as an electron donor and one or more of its organic products acts as a final electron acceptor.

          A.   Antibiotics

               Raw materials containing sugar are fermented by both eukaryotes and prokaryotes producing antibiotics as end products.

          B.   Yogurt

               Milk sugar (lactose) is broken down by bacteria via the Embden-Meyerhof pathway to lactate, formate, and acetate with small amounts of acetaldehyde and diacetyl, depending on the species of bacteria used.

          C.   Kombucha tea

               Tea with sucrose added is broken down by a "kombucha mushroom," consisting of many different cultures of yeasts and bacteria, to many different products, e.g., lactic acid, acetic acid.

          D.   Sauerkraut and kim chi

               Sugar in cabbage is broken down into lactic acid by different bacteria.

          E.   Beer

               Wort (sugar from grains) is broken down by yeast into ethanol.

          F.   Root beer/birch beer

               Plant and root extracts along with sucrose are broken down by yeast.

Laboratory #6
Microbial Observation

Name:_____
Date:_____

General physical description of the medium and where used

Broth Medium	**Agar Slant**	**Agar Deep**	**Agar Plate**
Describe:	Describe:	Describe:	Describe:
_____	_____	_____	_____
_____	_____	_____	_____
_____	_____	_____	_____
_____	_____	_____	_____
_____	_____	_____	_____
_____	_____	_____	_____
_____	_____	_____	_____

1. Why is aseptic technique important?

2. What are differences between respiration and fermentation?

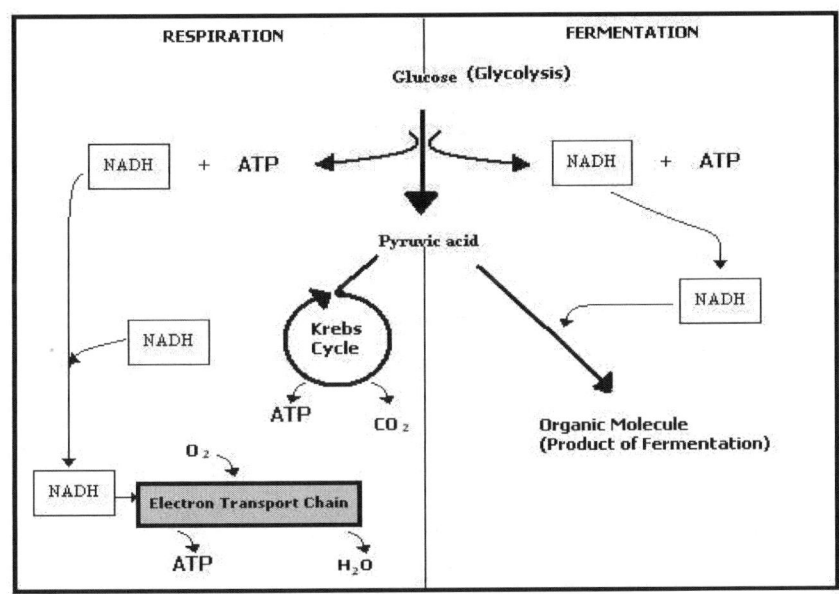

FERMENTATION LAB

See supplemental labs in back of book, depending on what activity is to be performed

Sauerkraut/Kim Chi
Yogurt
Kombucha
Beer
Root Beer/Birch Beer

Laboratory #7
Microbes in the Environment and Changing the Environments

Germs may infect those who live with persons infected, and germs can be preserved for a certain time, not only in fomes but also in the air.
Girolamo Fracastoro 1546

Part I: Environment

Introduction: Microbes are considered ubiquitous organisms or, more specifically, they are everywhere; they are found in the water we drink, the air we breathe, and the earth we walk on. Most of these microorganisms are non-pathogenic and present no problem to the healthy individual. There are, however, instances when **transient** pathogenic organisms get transferred from a **reservoir** of infection to a susceptible host. Infectious diseases remain the number one killer worldwide, but thanks to advanced medical care and public sanitation, infectious diseases run a close third to heart disease (number one killer) and cancers (number two killer) in the United States and other developed countries.[1] Latest evidence implicates infectious germs as a trigger for many cases of heart disease and many kinds of cancer. When researchers put cameras in public rest rooms to track people's behavior, the percentage of those who do not properly wash their hands, or do not wash at all is staggering, well over 60%.[2] There are 2.5 million **nosocomial** infections every year, which cause 30,000 deaths and contribute substantially to another 90,000 deaths.[3] Most of these are 100% preventable and should make anyone stop and think before going into a hospital for any period, however brief. Eighty percent of these infectious disease are transmitted by touch either **directly**, by **contact** with another person, or **indirectly**, by **contact** with something that person has touched.[2] We are about to see where these germs are found.

I.   Microbes in the environment - ubiquitous
    A.   Microbes are everywhere
        1.   In the water we drink
        2.   In the air we breathe
        3.   The earth we walk on
        4.   They live in and on our bodies
    B.   Most of the time they are harmless
II.   Culturing these microorganisms The Great Plate Count Anomaly
    A.   **Chemically defined medium**
        Exact chemical composition known
    B.   **Complex medium**
        Exact chemical composition varies slightly from batch to batch
III. Nutrient Agar
    A.   Commonly used medium to grow a wide variety of organisms
    B.   Organic carbon, energy, and nitrogen sources usually supplied by protein in the form of:
        1.   Meat extracts (beef extract)
        2.   Partially digested proteins (peptones)
    C.   Nutrient agar
        Ingredients per 100ml reconstituted media
        Peptone   0.5g
        Beef extract   0.3g
        Sodium chloride 0.8g
        Agar   1.5g

Distilled water   100ml

    D.   Preparation
- 1. Weigh out ingredients and add water
- 2. Heat until boiling
- 3. Sterilize via steam sterilization 121^0 C for 15'
- 4. Pour aseptically 15-20ml per 100mm X 15mm petri dish

    E.   **RODAC** plates (Replicate Organism Detection and Counting) will be demonstrated and also will be available to use.

    F.   **Incubation** - suitable conditions for bacteria to grow (multiply).

    G.   **Colony** - a population of cells arising from a single bacterial cell. A colony may arise from a group of the same microbes attached to one another, therefore called a **colony forming unit (cfu)**.

IV.   Lab Exercise

    A.   Design your own experiment. The purpose is to sample your environment. Use your imagination. Each group will have 2 sterile nutrient agar plates, 1 sterile RODAC plate, swabs, and sterile saline to moisten the swabs. Here are some suggestions for areas to sample:

- 1. You may use the lab, a washroom, or any place in the building for the environment.
- 2. Inoculate a plate from an environmental surface such as the floor or workbench by wetting a sterile swab (cotton) in sterile media/water-- swabbing the surface then swabbing the agar surface. Procedure to be demonstrated.
- 3. RODAC contact plates can be used to sample hard, flat surfaces Procedure to be demonstrated.

    B.   Incubate the plates inverted in the incubator – 35^0 C.

    C.   I will remove plates after 48 hours and refrigerate.

    D.   Next time we will observe the growing organisms.

Part II: Changing Environment

Introduction: The presence or absence of molecular oxygen can be very important to the growth of bacteria. Some bacteria, called **obligate aerobes**, require oxygen, while others called, **obligate anaerobes**, unable to tolerate oxygen. One reason obligate anaerobes cannot tolerate the presence of oxygen is that they lack catalase/peroxidase and the resultant accumulation of hydrogen peroxide is lethal. **Aerotolerant anaerobes** cannot use oxygen but tolerate it fairly well (fermentative bacteria). **Microaerophiles** grow best at increased carbon dioxide (5-10%). **Capnophiles** grow best at concentrations of carbon dioxide 10 to 30 times higher than in the atmosphere and are often found in host tissues supporting this concentration. The majority of bacteria are **facultative anaerobes**, which are capable of living with or without oxygen. Therefore to culture these different types of organisms different methods of incubation are required and these will be demonstrated.

V.   Discussion of Incubation Parameters - Specifically Special Atmospheric Conditions Different than Aerobic

    A.   The challenges
- 1. Obligate anaerobes
  Example: *Clostridium spp.*

        2.   Capnophiles

             Example: *Neisseria spp.*

        3.   Microaerophiles

             Example: *Streptococci spp.*

B.   Changing Atmospheric Conditions

        1.   **Reducing media** - drive out oxygen by boiling for 10 minutes

        2.   **Reducing media** - bind any oxygen diffusing in by adding reducing agents

           a.   Sodium thioglycollate

           b.   Oxyrase - enzyme which breaks down oxygen

        3.   **Carbon dioxide incubator** - microaerophiles

        4.   **Candle jar** - microaerophiles and capnophiles

        5.   **GasPak jar** - anaerobes

[1]Centers for Disease Control and Prevention (2009). *Health, United States, 2009.* Retrieved June 20, 2010 from www.cdc.gov/nchs/data/hus/hus09.pdf

[2]Tierno, P. (2001). *The Secret Life of Germs: Observations & Lessons from a Microbe Hunter.* New York, NY: Atria Books.

[3]Nguyen, Q. (2009). *Hospital-Acquired Infections.* Retrieved March 30, 2010, from http://emedicine.medscape.com/article/967022-overview.

Laboratory #7
Microbial Observation

Name: _____

Date: _____

Nutrient Plate #1
Describe location briefly:

Number of organisms_____
Types of organisms_____

Nutrient Plate #2
Describe location briefly:

Number of organisms_____
Types of organisms_____

RODAC Plate #1
Describe location briefly:
Number of organisms_____
Types of organisms_____

1.   Why is the swab moistened with a sterile liquid prior to swabbing?

2.   The inoculated agar plate is incubated inverted to allow for condensation to collect on the lid. What causes the condensation? Why is condensation on the agar undesirable?

3 .   Will all the organisms living in or on the area sampled grow on your nutrient agar plates? Briefly explain.

Demonstration of Oxygen Requirements and Corresponding Growth
(Instructor will Demonstrate)

Thioglycollate Medium for Anaerobes: *Clostridium perfringens*   **Anaerobe**

Gas Pak Anaerobic Chamber for Anaerobes: *Clostridium perfringens*   **Anaerobe**

Candle Jar for Capnophiles: *Streptococcus pneumoniae* **Capnophile**

Aerobic Air Incubator for Aerobes   *Bacillus subtilis*   **Aerobe**
and Facultative Anaerobes: *Pseudomonas aeruginosa* **Aerobe**
          Escherichia coli   **Facultative**
          Staphylococcus aureus   **Facultative**

Laboratory #8
Beginning Enzyme Profiles

To say that a man is made up of certain chemical elements is a satisfactory description only for those who intend to use him as a fertilizer.
-**Hermann Joseph Muller**

Introduction: The chemical reactions observed by Pasteur and the chemical reactions that occur within all living organisms are referred to as **metabolism**. Metabolic processes involve **enzymes**, which are proteins that catalyze biological reactions. The majority of enzymes function inside a cell - that is, they are endoenzymes. Many bacteria make some enzymes, called exoenzymes, that are released from the cell to catalyze reactions outside of the cell. Because many bacteria share the same colonial and cell morphology, additional factors, such as metabolism, are used to identify them. This exercise will provide a "whetting of the appetite, if you will," for the beginning of bacteria identification.

I.   **Qualitative** analysis - identification of Bacteria
    A.   Staining-direct examination
        1.   Simple stain
        2.   Differential stain (gram stain, acid-fast stain)
        3.   Structural stains (spore stain, capsule stain, flagella stain)
    B.   DNA probes
        1.   Very good but expensive and requires techniques
        2.   **PCR** - polymerase chain reaction - small samples of DNA quickly amplified to quantities large enough to analyze - uses DNA polymerase and adds four nucleotides to small sections of sample DNA to make multiple copies.
    C.   Enzyme profiles or biochemical analysis
        1.   Tried and true methodology
        2.   Fairly inexpensive
        3.   Many test kits available
        4.   Based on enzymes present in certain bacteria, which react to certain added substrates (substrates are the bait and the enzymes the predator).
II.   Carbohydrate Metabolism
    A.   Most bacteria catabolize carbohydrates for energy and carbon source
    B.   **Exoenzymes** - hydrolytic enzymes which leave the cell and break down large substrates into smaller usable substrates via hydrolysis
    C.   Smaller substrates can then be utilized or catabolized further, e.g., glucose
        1.   Some catabolize oxidatively, producing carbon dioxide and water
        2.   *Most* ferment without using oxygen
    D.   Laboratory determination if oxidative or fermentative
        1.   OF media
            a)   Semi-solid agar deep containing a high concentration of carbohydrate and a low concentration of peptone (protein digest) with bromthymol blue as the indicator.
            b)   Two tubes used - one open to the air and the other covered with sterile mineral oil.
            c)   If both turn yellow (acid production final breakdown) then carbohydrate is fermented.
            d)   If only the open tube turns yellow then it is oxidative.
            e)

III. **Starch Hydrolysis**

    A.   If organism produces enzyme amylase to break down starch, amylase can be detected by an iodine test.

    B.   BPS (breakdown products of starch) - amylase and maltase will break down polymers to form:

        1.   Glucose - monosaccharide

        2.   Maltose - disaccharide

        3.   Dextrin - polysaccharide

    C.   Iodine will turn starch blue if present (potato trick) - if starch not present it will remain yellow.

    D.   Divide starch agar plate into two sections and label bottom of plate.

    E.   Streak short single lines of *Bacillus spp.* and *Escherichia coli.*

    F.   Incubate 24 hours in incubator.

    G.   Next class - flood plate with iodine and read reactions - fill out chart and answer questions -- *iodine will turn purple in presence of starch.*

IV.   **Catalase** test for Presence of Enzyme Catalase

    A.   During aerobic respiration hydrogen atoms may combine with oxygen and form hydrogen peroxide.

    B.   Hydrogen peroxide is toxic to cells, so many organisms overcome this by producing enzyme catalase, which breaks down hydrogen peroxide to water and oxygen (*bubbles*).

$$2H_2O_2 ----------catalase----------- 2H_2O + O_2$$

    C.   Very important in clinical lab–separate the Staphs from the Streps.

    D.   Divide tryptic soy agar plate in half and mark bottom of plate.

    E.   Streak short single line of *Strep* on one side and *Staph* on the other side.

    F.   Incubate 24 hours in incubator. Next class - drop 3% hydrogen peroxide on streak pattern and observe for bubbling or oxygen release.

    G.   Fill out chart and answer questions.

V.   **Mannitol salt agar** - a selective agar for growing only halodures (>7.5% salt) and a differential agar for differentiating mannitol fermenters.

    A.   Halodures are salt tolerant bacteria found on the skin - Staph, *Micrococcus.*

    B.   Divide mannitol salt agar plate into four sections and label bottom of plate.

        1.   *Staphylococcus epidermidis*

        2.   *Staphylococcus aureus*

        3.   Touch one partner's two fingers

        4.   Touch other partner's two fingers

    C.   Incubate 24 hours in incubator. Next class - observe and fill out chart and answer questions.

    D.   Mannitol fermenters produce lactic acid as a by-product, which changes the agar to a yellow color.

VI.   **MacConkey agar** - a selective agar for growing only gram-negative organisms and a differential agar for differentiating lactose fermenters from non-lactose fermenters

    A.   Bile salts and crystal violet dye inhibit gram-positive organisms.

    B.   Lactose fermenters will be red opaque colonies because acids of lactose fermentation will act upon the bile salts and absorb the neutral red dye--non-lactose fermenters produce colorless translucent colonies.

    C.   Divide plate into three sections and label bottom of plate.

    D.   Streak *Escherichia coli, Salmonella spp., & Staphylococcus aureus.*

E.   Incubate 24 hours in incubator.

F.   Next class-observe and fill out chart and answer questions.

VII.   **Oxidase Test** - an enzyme used to differentiate a select few microorganisms which have cytochrome c in the electron transport chain.

    A.   *Pseudomonas spp.*

    B.   *Neisseria spp.*

    C.   *Alcaligines spp.*

    D.   Place filter paper in petri dish.

    E.   Crush an oxidase ampoule and drop on filter paper.

    F.   Immediately rub a loop of *Pseudomonas aeruginosa* from an agar plate and not from broth over top of moistened paper.

    G.   Immediately rub a loop of *Escherichia coli* over top of moistened paper.

    H.   Observe color change

        1.   Blue to black = positive for oxidase

        2.   No change = negative for oxidase

Laboratory #8
Microbial Observation

Name:_____

Date:_____

STARCH HYDROLYSIS

Organism	Growth	Reaction
Staphylococcus aureus		
Bacillus subtilis		

CATALASE TEST

Organism	Appearance after H_2O_2	Catalase reaction
Streptococcus spp.		
Staphylococcus spp.		

MANNITOL SALT AGAR

Organism	Growth	Reaction
Staphylococcus aureus		
Staphylococcus epidermidis		
Fingers – Lab Partner #1		
Fingers – Lab Partner #2		

MacCONKEY AGAR

Organism	Growth	Reaction
Staphylococcus aureus		
Escherichia coli		
Salmonella spp.		

CYTOCHROME OXIDASE

Organism	Growth	Reaction
Pseudomonas aeruginosa		
Escherichia coli		

1. Which organism gave a positive test for starch hydrolysis? How can you tell?

2. Why is it important to first determine if growth occurred in a differential media such as starch agar, before examining the plate for starch hydrolysis?

3. Why does hydrogen peroxide bubble when it is poured on a skin cut?

4. Would you think the mannitol salt agar would be the only test used to determine if pathogenic Staph are present?

5. What type of media is MacConkey agar? Mannitol salt agar? Tryptic soy agar?

Laboratory #9
Microbial Flora of the Human Body
And Effectiveness of Hand Washing – Supplemental Lab

There are always germs, and you're never going to get rid of all of them.

Cheryl Mendelson

Introduction:

The respiratory tract can be divided into two systems: the upper and lower respiratory systems. The **upper respiratory system** consists of the nose and throat, and **the lower respiratory system** consists of the larynx, trachea, bronchial tubes, and alveoli. The lower respiratory tract is normally sterile because of the efficient functioning of the **ciliary escalator**. The upper respiratory system is in contact with the air we breathe--air contaminated with microorganisms.

The throat is a moist, warm environment, allowing many bacteria to establish residence. Species of many different genera--such as *Staphylococcus, Streptococcus, Neisseria,* and *Haemophilus*-- can be found living as normal microbiota in the throat. Despite the presence of potentially pathogenic bacteria in the upper respiratory system, the rate of infection is minimized by **microbial antagonism**. Certain microorganisms of the normal microbiota suppress the growth of other microorganisms through competition for nutrients and production of inhibitory substances.

Streptococcal species are the predominant organisms in throat cultures, and some species are the major cause of bacterial sore throats (acute pharyngitis). *Streptococci* are identified by biochemical characteristics, including hemolytic reactions, and antigenic characteristics (Lancefield's system). Hemolytic reactions are based on hemolysins that are produced by *Streptococci* while growing on blood enriched agar. Blood agar is usually made with defibrinated sheep blood (5.0%), sodium chloride (0.5%) to minimize spontaneous hemolysis, and nutrient agar. Three patterns of hemolysis can occur on blood agar:

1. **Beta hemolysis**: Complete hemolysis, producing a clear zone and clean edge around the colony
2. **Alpha hemolysis**: Incomplete hemolysis, producing methemeglobin and a green, cloudy zone around the colony
3. **Gamma hemolysis**: No hemolysis, producing no change in the blood agar around the colony

Streptococci that are alpha hemolytic and gamma hemolytic are usually the normal microbiota, whereas **beta hemolytic** *Streptococci* are frequently pathogens.

The *Streptococci* can be antigenically classified into Lancefield groups A through H and K through V, by group specific carbohydrates or antigens in their cell walls. About 25% of sore throats are caused by **beta hemolytic group A** *Streptococci*.[1] These bacteria are assigned to the species *Streptococcus pyogenes* and are sensitive to the antibiotic Bacitracin; other *Streptococci* are resistant to Bacitracin. *Streptococcus pneumoniae* is alpha hemolytic and a pathogen and is resistant to Optochin, another antibiotic. These antibiotics are also used to differentiate the *Streptococci*.

38

The nasal cavity is another warm, moist environment conducive to microbial growth. Approximately 10-40% of the population harbors *Staphylococcus aureus*, without causing any complications; however, under certain circumstances these same organisms can lead to infection.[2]

The resident micro flora of the skin is predominantly *Staphylococcus epidermidis*, with some Streptococcal species, some yeasts, and some fungi also present at certain times.

Procedure (throat culture):

1. Swab your throat with a sterile cotton swab, which does not need to be moistened. The area to be swabbed is between the "golden arches" (glossopalatine arch). SWAB THE TONSILS IF PRESENT AND DO NOT HIT THE TONGUE OR OTHER AREAS IN THE MOUTH.
2. After obtaining a specimen from the throat, swab one half of a blood agar plate. Discard the swab in biohazard container.
3. If strep throat is suspected place an A disk on the heaviest part of streak.
4. Incubate the plate inverted at 35-37^0 C for 24 hours.
5. Next lab: observe for hemolysis.

Procedure (nasal swab):

1. Swab the inside nares of the nose with a sterile cotton swab--swab does not need to be moistened.
2. Swab a mannitol salt agar plate and incubate inverted at 35-37^0 C for 24 hours.

Procedure (skin swab):

1. Swab a warm moist area of the skin, e.g., under the arm or axilla.
2. Swab this on a mannitol salt agar plate.
3. Incubate the mannitol salt agar plate inverted at 35-37^0 C for 24 hours.

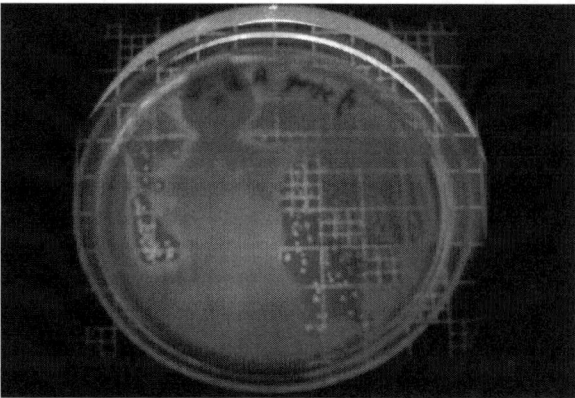

Throat Swab   Beta hemolytic strep with group A disk (bacitracin) = Group A Strep (GAS)

Procedure (Handwashing): *Refer to supplemental labs*

[1]Pediatric Dental Health. (2005). Retrieved March 30, 2010 from http://dentalresource.org/topic56scarletfever.html.
[2]Baron, S. (1996). *Medical Microbiology 4th ed.* Retrieved March 30, 2010 from http://www.ncbi.nlm.nih.gov/bookshelf/br.fcgi?book=mmed&part=A512.

Laboratory #9
Microbial Observation

Name:_____

Date:_____

Throat Culture

Test	Appearance on Blood Agar
Hemolysis	
A Disk (*if used*)	

1. Why must you transfer part of colony to a slide instead of performing the catalase reaction on a blood agar plate?

2. Is blood agar selective or differential? Briefly explain.

3. You have isolated a gram positive coccus from a throat culture that you cannot identify as *Staphylococcus* or *Streptococcus*. A test for one enzyme can be used to distinguish quickly between these bacteria. What is that enzyme?

Nose and Skin Culture

Body Site	Mannitol Salt Agar Reaction	Presumptive Identification
Nose		
Skin		

1. What are some skin infections caused by *Staphylococci*?

2. What are some skin infections caused by *Streptococci*?

3. Are open wound infections, like impetigo, contagious? If yes, how could they be spread?

Laboratory #10
Microbes in Food & Water
Standard Plate Count & Membrane Filtration

Red meat is not bad for you. Now blue-green meat, that's bad for you.

Tommy Smothers

Introduction:

Illness and food spoilage can result from microbial growth in foods. The sanitary control of food quality is concerned with the testing of food for the presence of pathogens. During processing (grinding, washing, packaging, etc.), food may become contaminated with soil microbes and microbiota from animals, food handlers, and machinery. Foods are the primary vehicle responsible for the transmission of diseases of the digestive system. For this reason, they are examined for the presence of coliform bacteria, indicative of fecal contamination, and for the presence of heterotrophic bacteria, tested by the **standard plate count**. The standard plate count is used to determine the total number of *viable* bacteria in a food sample. The presence of large numbers of bacteria is undesirable in most foods because it increases the likelihood that pathogens will also be present and thus increases the potential for food spoilage. In a standard plate count, the total number of colony forming units (cfu) is determined. Each colony may arise from a group of cells rather than from one individual cell. The initial sample is diluted through serial dilutions in order to obtain a small number of colonies on each plate. A known volume of the diluted sample is placed in a sterile petri plate and melted--cooled nutrient agar is poured on the sample, after incubation, the colonies are counted. Plates with between 25 and 250 colonies are suitable for counting. The microbial population can then be calculated using the following equation:

$$\textbf{CFU/g or ml sample} = \frac{\textbf{Number of colonies}}{\textbf{Quantity plated}} \times \textbf{Dilution}$$

Procedure (bacteriological examination of milk):

1. Obtain a sample of either pasteurized or raw milk.

2. Using a sterile 1ml pipette, aseptically transfer 1ml of the milk sample into a 9ml saline blank tube; label 10^{-1} and discard the pipette. Shake the tube 20 times with your elbow resting on the table as shown in the diagram.

3. Using a sterile 1ml pipette, aseptically transfer 1ml of the 10^{-1} milk sample into a 9ml saline blank tube; label 10^{-2} and discard the pipette. Shake the tube 20 times with your elbow resting on the table.

4. Using a sterile 1ml pipette, aseptically transfer 1ml of the 10^{-2} milk sample into a 9ml saline blank tube; label 10^{-3} and discard the pipette. Shake the tube 20 times with your elbow resting on the table.

5. Label the bottoms of four petri dishes with the dilutions 10 X, 10^2 X, 10^3 X, and 10^4 X. *Note: Ideally, this should be done in duplicate for statistical accuracy.*

6. Transfer 1ml from each dilution in the designated tubes into the bottom of the empty sterile petri dishes. For the 10^4 X, transfer 0.1ml from the 10^{-3} dilution.

7.   Check the temperature of the melted nutrient agar with the back of your hand or your cheek (baby bottle warm). Pour the melted nutrient agar into each of the plates (about one third full or approximately 15ml) and swirl plate carefully.

8.   When the agar has solidified, invert and incubate at 35° C for 48-72 hours.

Procedure (bacteriological examination of hamburger):

1.   5g of hamburger is blended in a sterile blender with 500ml of sterile buffer. This 1-100 dilution will be prepared for you by the instructor.

2.   Using a sterile 1ml pipette, aseptically transfer 1ml of the 10^{-2} hamburger-saline solution into a 9ml saline blank tube; label 10^{-3} and discard the pipette. Shake the tube 20 times with your elbow resting on the table as shown in the diagram.

3.   Using a sterile 1ml pipette, aseptically transfer 1ml of the 10^{-3} sample into a 9ml saline blank tube; label 10^{-4} and discard the pipette. Shake the tube 20 times with your elbow resting on the table.

4.   Using a sterile 1ml pipette, aseptically transfer 1ml of the 10^{-4} sample into a 9ml saline blank tube; label 10^{-5} and discard the pipette. Shake the tube 20 times with your elbow resting on the table.

5.   Using a sterile 1ml pipette, aseptically transfer 1ml of the 10^{-5} sample into a 9ml saline blank tube; label 10^{-6} and discard the pipette. Shake the tube 20 times with your elbow resting on the table.

6.   Label the bottoms of four petri dishes with the dilutions 10^3 X, 10^4 X, 10^5 X, and 10^6 X. *Note: Ideally, this should be done in duplicate for statistical accuracy.*

7.   Transfer 1ml from each dilution, beginning with 10^{-3}, into the bottom of the empty sterile petri dishes.

8.   Check the temperature of the melted nutrient agar with the back of your hand or your cheek (baby bottle warm). Pour the melted nutrient agar into the plates (about one third full or approximately 15ml) and swirl plate carefully.

9.   When the agar has solidified, invert and incubate at 35° C for 48-72 hours.

Counting colonies next class period--SEE SUPPLEMENTAL LAB 1J FOR COUNTING

1.   Arrange each plate in order from lowest to highest.

2.   Select the plate with 25 to 250 colonies (most accurate). We will count all colonies.

3.   Calculate the number of bacteria in the original food. For example if 129 colonies were counted on a 10^3 X dilution: *129 colonies* = 129/1ml X 10^3 cfu/g (ml) or 129,000 cfu/g (ml)

Laboratory #10
Microbial Observation

Name: _____

Date: _____

SAMPLE: _____ **Milk** _____

Dilution	Colonies per Plate

Number of colony forming units per ml of original milk sample_____
(calculations):

SAMPLE: _____ **Hamburger** _____

Dilution	Colonies per Plate

Number of colony forming units per g of original food sample_____
(calculations):

Membrane Filter Method for Coliform Analysis of Drinking Water—Demonstration

1. Why are plates with 25 to 250 colonies used for calculations and others discarded?

2. In a quality control laboratory, each dilution is plated in duplicate or triplicate. Why would this increase the accuracy of a standard plate count?

3. Why is ground beef a better bacterial growth medium than a steak or roast?

4. Why does repeated freezing and thawing increase bacterial growth in meat?

Laboratory #11
Chemical Methods of Control
Antiseptics, Disinfectants, & Antibiotics

Antibiotics are truly miracle drugs that have saved countless millions of lives. But antibiotic resistance is a critical public health issue that is eroding the effectiveness of antibiotics and may affect the health of each and every one of us.

Betsy Bauman

Introduction Disinfectants and Antiseptics - Use-Dilution Test (-cidal): Disinfectants are chemical agents used on inanimate objects to lower the level of microbes on their surface; **antiseptics** are chemicals used on living tissue to decrease the number of microbes. Disinfectants and antiseptics affect bacteria in many ways. Those that result in bacterial death are called **bacteriocidal** agents. Those causing temporary inhibition of growth are **bacteriostatic** agents. No single chemical is the best to use in all situations. Antimicrobial agents must be matched to specific organisms and environmental conditions. The standard method for evaluating the effectiveness of a chemical agent is the **American Official Analytical Chemists use-dilution test.** For most purposes, three strains of bacteria are used in the test: *Salmonella cholerasius, Staphylococcus aureus*, and *Pseudomonas aeruginosa*. In this first exercise we will perform a modified use-dilution test.

Procedure:
1. We will use *Staphylococcus aureus* as our culture of choice.
2. Prepare a dilution of the test substance, diluted to the strength normally used.
3. We will use concentrations of isopropyl alcohol, e.g., 10%, 25%, 50%, 70%, 90%.
4. Transfer 5 ml of this diluted suspension to a sterile test tube.
5. Divide 1 tryptic soy agar plate into 4 sections and label 0, 30", 60", 5'.
6. Inoculate or streak 0 section with a loop of *Staphylococcus aureus* culture (control).
7. Aseptically add approximately 15 drops of *S. aureus* to the diluted test substance and mix.
8. Transfer 1 loop from the mixed suspension to the corresponding sections at 30", 60", and 5'.
9. Incubate the plates inverted at 35-37°C for 24 hours.

Introduction Zone of Inhibition Testing for Disinfectants and Antiseptics (-static): Agar plates are inoculated uniformly with the desired organism (5ml of 24hr culture to 100ml tempered agar). Plugs are withdrawn from the inoculated plates, creating wells to add tested substances. The substances being evaluated diffuse through the inoculated agar from an area of high concentration to an area of lower concentration. An effective agent will inhibit bacterial growth, and measurements can be made of the size of the zones of inhibition around the wells. The concentration of chemotherapeutic agent at the edge of the zone of inhibition represents the **minimum inhibitory concentration (MIC)**. The MIC is determined by comparing MIC values with values from a standard table. The zone size is affected by such factors as diffusion rate of the substance and growth rate of the organism used.

Procedure:
1. We will use *Staphylococcus aureus* and/or *Pseudomonas aeruginosa* as the organisms of choice.
2. 0.5 ml of the culture was added to 7 ml of melted TSA agar and poured over top of a TSA plate.

3.　　A plug of agar was removed from four areas of each plate and sealed with one drop of melted agar.

4.　　Fill the empty wells with the corresponding disinfectant, being sure to label and incubate at 35-37^0 C for 24 hours. DO NOT INVERT.

Introduction to Antibiotics: The observation that some microbes inhibited the growth of others was made as early as 1874. Pasteur and others observed that infecting an animal with *Pseudomonas aeruginosa* protected the animal against *Bacillus anthracis*. Later investigators coined the word antibiosis (against life) for this inhibition and called the inhibiting substance an antibiotic. In 1928, Alexander Fleming observed antibiosis around a mold (*Penicillium*) growth on a culture of *Staphylococci*. He found that culture filtrates of *Penicillium* inhibited the growth of many gram-positive cocci and also *Neisseria spp*. In 1940, Selman Waksman isolated the antibiotic streptomycin, produced by an actinomycete, or soil bacteria. This antibiotic was effective against many bacteria that were not affected by penicillin. Actinomycetes remain an important source of antibiotics to this day. Antimicrobial chemicals absorbed or used internally, whether natural (antibiotics) or synthetic, are called chemotherapeutic agents. A physician or dentist needs to select the correct chemotherapeutic agent intelligently and administer the appropriate dose in order to treat an infectious disease; then the practitioner must follow that treatment in order to be aware of resistant forms of the organism that might occur. The clinical laboratory isolates the pathogen from a clinical sample and determines its sensitivity to chemotherapeutic agents. In the **disk-diffusion method**, a petri dish containing an agar growth medium is inoculated uniformly over its entire surface with the culture to be tested. Paper disks impregnated with various chemotherapeutic agents are placed on the surface of the agar. During incubation, the chemotherapeutic agent diffuses off the disk, from an area of *higher concentration to an area of lower concentration*. An effective agent will inhibit bacterial growth, and measurements can be made of the size of the zones of inhibition around the disks. The concentration of chemotherapeutic agent at the edge of the zone of inhibition represents its minimum inhibitory concentration (MIC). The MIC is determined by comparing the zone of inhibition with MIC values in a standard table. Zone size is affected by: (1) diffusion rate and (2) growth rate of the organism. To minimize the variance between laboratories, the standardized **Kirby-Bauer test** for agar diffusion methods is performed in many clinical laboratories, following strict quality controls. This test uses **Mueller-Hinton agar**, which allows the chemotherapeutic agent to diffuse freely and is standardized worldwide.

Materials:

> Petri dish containing mueller-hinton agar
> Culture in tryptic soy broth
>> *Staphylococcus aureus,* or
>> *Escherichia coli*
> Sterile swabs
> Forceps
> Gram-positive antimicrobial disks, or
> Gram-negative antimicrobial disks

Antibiotic Susceptibility Testing Procedure

1. Aseptically swab the assigned culture onto the appropriate mueller-hinton agar plate. Swab in 3 directions to ensure complete plate coverage, by rotating the plate 90° each time. Let stand at least 5 minutes

2. Place appropriate disks on the inoculated plate by dropping on with a forceps, which has been dipped in alcohol and then flamed to dry the alcohol. Tap the disk with gentle pressure using a flame sterilized loop.

3. Place 4 disks appropriately spaced on the plate, equally spaced. Use 1 extra disk, a penicillin disk, if using *E. coli*, placing the penicillin disk in the center of the inoculated plate.

4. Record the agents and the disk codes in your laboratory report and circle the corresponding codes on the table attached.

5. Incubate the plates inverted, at 35-37° C for 24 hours.

Measuring Zones of Inhibition (MIC) next class period

1. Measure the zones of inhibition in millimeters, using a ruler on the underside of the plate.

2. Record the zone size and, based on the values from the table, indicate whether the organism is sensitive, intermediate, or resistant.

Interpretation of Inhibition Zones of Test Cultures[1]

Antimicrobial Agent	Disk Content	Resistant mm or less	Intermediate mm range	Susceptible mm or more
Chloramphenicol	30mcg	12	13-17	18
Erythromycin	15mcg	13	14-17	18
Kanamycin	30mcg	13	14-17	18
Naladixic Acid	30mcg	13	14-18	19
Nitrofurantoin	300mcg	14	15-16	17
Novobiocin	30mcg	17	18-21	22
Penicillin	10iu	28	-------	29
Streptomycin	10mcg	11	12-14	15
Tetracycline	30mcg	14	15-18	19
Sulfonamides (Triple)	250mcg	12	13-16	17

Gram-positive discs

Erythromycin (E)   Penicillin (P)
Chloramphenicol (C)   Novobiocin (N)
Tetracycline (T)   Streptomycin (S)

Gram-negative discs

Kanamycin (K)   Nitrofurantoin (F/M)
Chloramphenicol (C)   Nalidixic Acid (NA)
Triple Sulfa (SSS)   Tetracycline (T)

[1]Cappuchino, C. & Sherman. N. (2002). *Microbiology A Laboratory Manual 6th edition.* San Francisco, CA: Benjamin Cummings

Laboratory #11-Antiseptics & Disinfectants
Microbial Observation

Name:_____

Date:_____

Disinfectants and Antiseptics
Antiseptic Activity – Use-Dilution

Time of Exposure	Control-*S aureus*	Sanitizer #1 (Percent Alcohol)
0		-----------------------
30 seconds	-----------------------	
60 seconds	-----------------------	
5 minutes	-----------------------	

Disinfectants and Antiseptics
Zone of Inhibition

Disinfectant/Antiseptic	*Pseudomonas aeruginosa* zones	*Staphylococcus aureus* zones

Antibiotic Testing

Antimicrobial Agent	Code	*S. aureus* zone	*S. aureus* S, I, or R	*E. coli* zone	*E. coli* S, I, or R

S = Susceptible; I = Intermediate; R = Resistant

1. Which chemotherapeutic agent was the most effective against each organism?

2. Is the disk diffusion method measuring bacteriostatic or bacteriocidal activity? Briefly explain.

3. What is the minimum inhibitory concentration (MIC) of each antibiotic?

Laboratory #12
Unknown Bacterial Identification

Genius is one percent inspiration, ninety-nine percent perspiration.

Thomas A. Edison

Introduction: In microbiology, a system of classification must be available to allow the microbiologist to categorize and classify organisms. Communication among scientists would be very limited if no universal system of classification existed. Until recently, the taxonomy (grouping) of bacteria was difficult because few definite anatomical or visual differences exist. With these limitations, most bacteria are identified through evaluation of primary characteristics, such as morphology and growth patterns, and secondary characteristics, such as **metabolism** and serology.

The taxa used for the bacteria and archaea are phylum, class, order, family, genus, and species. Although the characteristics of a given group are relatively constant, through repeated laboratory culture, **atypical bacteria** will be found. This versatility, however, only heightens the fun of classifying bacteria.

You will be given an unknown, **chemoheterotrophic** bacteria to characterize and identify. By using careful deduction and by systematically compiling and analyzing data, you should be able to identify the bacterium.

A key will be attached. The key is an example of a dichotomous classification system, whereby a population is repeatedly divided into two parts until a description identifies a single member.

To begin, simultaneously ascertain the purity of stock culture you have been given and perform a **gram stain** *(note: we only have one period so this is why we perform simultaneously)*. Avoid contamination of your culture when performing any tests to prevent incorrect identities.

The following will be at your disposal. Tryptic soy agar plates, mannital salt agar plates, MacConkey agar plates, blood agar plates, cytochrome oxidase strips, hydrogen peroxide, and an API 20E commercially packaged kit for enterobacteriaceae identification, along with the necessary API chemicals.

Your organism will be one of the following:

Staphylococcus aureus	*Salmonella enterititis*	*Citrobacter freundii*
Staphylococcus epidermidis	*Serratia marcescens*	*Proteus mirabilis*
Streptococcus pyogenes	*Pseudomonas aeruginosa*	*Escherichia coli*
Streptococcus pneumoniae	*Enterobacter cloacae*	*Bacillus cereus*
Enterococcus faecalis	*Klebsiella pneumoniae*	*Salmonella typhi*

After you know the morphology and gram stain reactions of your culture, determine which of the above you will need to perform your identification. Use the key attached and all relevant data included with the API 20E Enterobacteriaceae kit to finalize your identification. With the identity of your organism finalized, write a one-page report about this organism. Include pathogenicity, clinical significance, nosocomial/iatrogenic significance, and any information you feel is relevant to this specific organism.

Procedure:

1. Streak unknown on a tryptic soy agar plate (TSA) for isolation and purity. Streak from Broth culture.

2. Perform gram stain from agar and broth culture.

3. Use, dichotomous key to determine which tests are then to be performed.

4. Gram positive organisms require catalase, mannitol salt agar, and blood agar if *Strep* is suspected.

5. Gram negative organisms require cytochrome oxidase, MacConkey agar, and API if organisms other than *Pseudomonas aeruginosa* are suspected.

6. Streak agar plates from broth culture.

7. Perform oxidase test using agar culture.

Results:

After 24 hours read the following tests: They may be read up to 72 hours, with the *Streptococcus spp.* much more visible after 48 hours.

1. Isolation plate is streaked to make sure you had a pure culture that was not contaminated with other organisms which would give erratic results. **Your instructor will read and assess this plate**.

2. Mannitol salt agar is used to determine if growth occurred at 7.5% salt concentration and if fermentation of mannitol occurred.

3. MacConkey agar is used to determine if organisms fermented lactose.

4. Blood agar is used to determine if *Streptococci* were alpha, beta, or gamma hemolytic.

5. API strips are used as per directions provided to speciate the gram negative organisms.

TAXONOMIC DICHOTOMOUS KEY
Identification of Medically Important Bacteria
Gram Stain Reaction
Gram Positive Bacteria

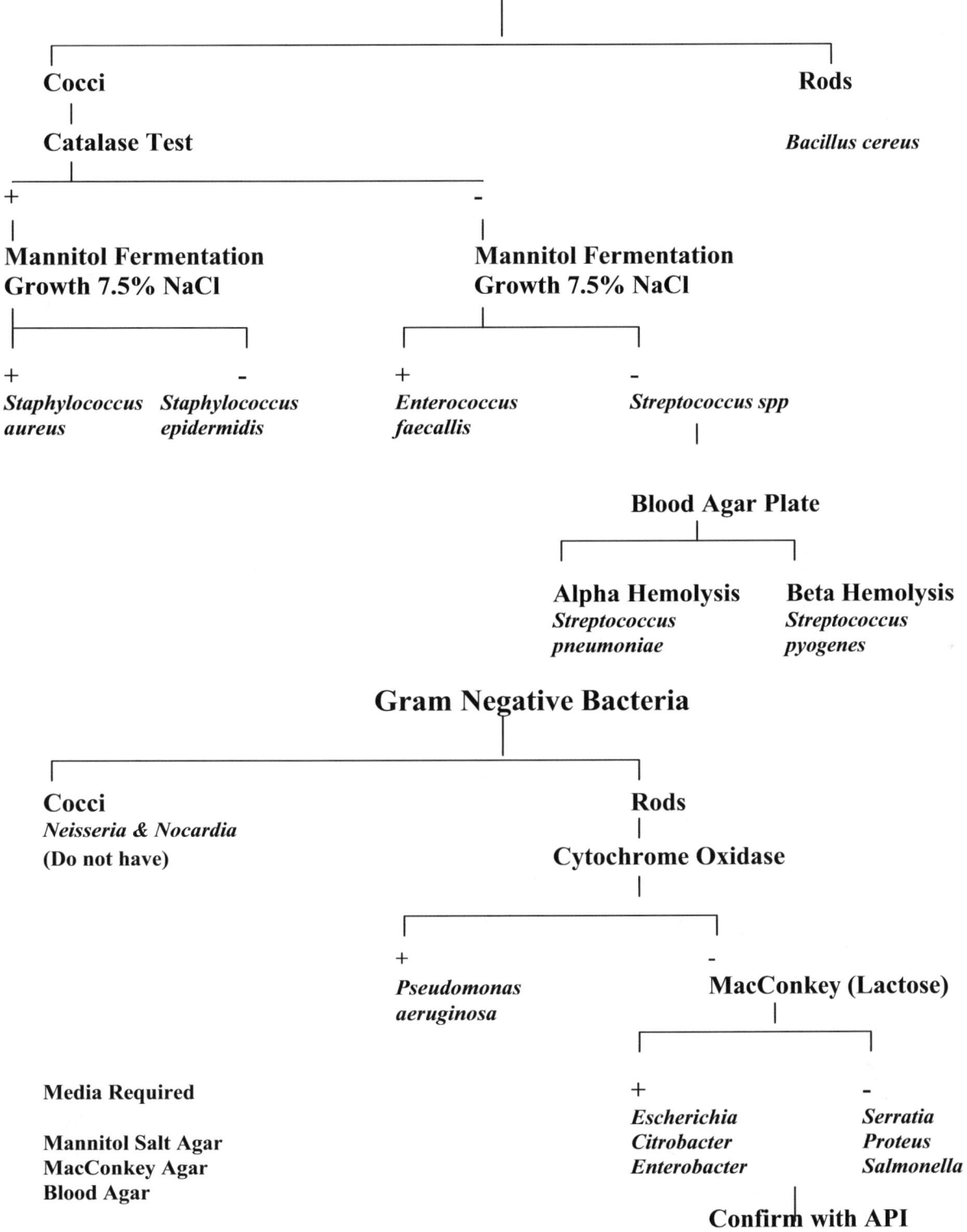

Cocci

Catalase Test

Rods

Bacillus cereus

+

Mannitol Fermentation
Growth 7.5% NaCl

-

Mannitol Fermentation
Growth 7.5% NaCl

+

Staphylococcus aureus

-

Staphylococcus epidermidis

+

Enterococcus faecallis

-

Streptococcus spp

Blood Agar Plate

Alpha Hemolysis
Streptococcus pneumoniae

Beta Hemolysis
Streptococcus pyogenes

Gram Negative Bacteria

Cocci
Neisseria & Nocardia
(Do not have)

Rods

Cytochrome Oxidase

+
Pseudomonas aeruginosa

-
MacConkey (Lactose)

Media Required

Mannitol Salt Agar
MacConkey Agar
Blood Agar

+
Escherichia
Citrobacter
Enterobacter

-
Serratia
Proteus
Salmonella

Confirm with API

API – 20E (bioMerieux Vitek, Inc.)[1]

A system composed of a plastic strip with 20 cupules containing dehydrated substrates or media. It is one of the most widely used biochemical identification systems commercially available.

Inoculation:

1. A well-isolated colony or more colonies, if a pure culture, is taken off the agar media and suspended in 5ml of 0.85% saline.

2. The suspension is mixed thoroughly, preferably by vortexing.

3. Using a sterile Pasteur pipet, inoculate the cupules as follows:

    A. All cupules are filled half-way except for **CIT**, **VP**, and **GEL**, which are completely filled (upper and lower).
    B. Fill by setting on bench and slightly tilting so inoculum flows into cup.

4. After all the cupules have been inoculated with the culture suspension, completely fill the cupules **ADH**, **LDC**, **ODC**, **H$_2$S**, and **URE** with sterile mineral oil.

5. Place the inoculated strip in the incubation tray and add about 3-5ml of water on the bottom, to create a humid environment.

6. Place the lid on the tray and incubate 18-24 hours at 35-37^{0}C. Longer incubation times are usually acceptable.

Interpretation of Results:

1. Read the reactions according to the chart supplied with the API kit. For further interpretation of reactions, refer to pictures on the API website.

2. Add necessary reagents, specified in API directions.

3. Interpret results on handout sheet by assigning corresponding numbers to the positive tests. Look up final profile index on the API website. This is a dynamic website, with profiles changed periodically as organisms undergo mutation.

---

[1]API Testing – Biomerieux. (n.d.). Retrieved March 19, 2010, from http://www.biomerieux-diagnostics.com/servlet/srt/bio/clinical-diagnostics/dynPage?doc=CNL_PRD_CPL_G_PRD_CLN_11

Laboratory #12
Microbial Observation

Name: _____

Date: _____

Culture Number: _____

Pure Culture: YES

List tests performed and agar streaked (how you arrived at your identification):

API-20E profile number (if applicable): _____

Organism identification: _____

Report on organism:

Include on separate sheets of paper

Supplemental Lab 1A
Genetics: Transformation: A Class Demonstration[1]

How many incomplete and false observations might have remained unpublished instead of swelling the bacterial literature into a turbid stream, if investigators had checked their preparations with each other?

Robert Koch

Introduction: Transfer of DNA into organisms has resulted in many advances in medicine, specifically human insulin and human growth hormones. It has also resulted in organisms becoming resistant to antibiotics. It is generally believed that organisms exchange DNA by one of three methods: (1) *transformation*, (2) *transduction*, or (3) *conjugation*. **Transformation**, the incorporation of naked DNA into competent bacteria, was first demonstrated by Frederick Griffith in 1928. He was able to demonstrate that solutions from killed pathogenic strains of *Streptococcus pneumoniae* would make otherwise non-pathogenic *Streptococcus pneumoniae* virulent or pathogenic. Griffith did not know what this transforming agent was, but 16 years later Colin MacLeod, Oswald Avery, and MaClyn McCarty, at the Rockefeller demonstrated that this agent was DNA. **Natural transformation** can occur in a few organisms (Institute for Medical Research), e.g., *Streptococcus pneumoniae*, *Neisseria gonorrhea*, *Haemophilus influenza*, and are not as rare as once thought, with virulence being passed between organisms. The DNA must first be transported into the recipient cell and then similarities have to exist on the recipient's DNA for the new DNA to be incorporated (recombination). **Artificial transformation** is accomplished by making recipient cells competent or making the cell membrane permeable to the foreign DNA.

Demonstration of Artificial Transformation:

Plasmid pVIB – genetically modified DNA containing the lux genes from *Vibrio fischeri* and a gene for resistance to the antibiotic ampicillin.

Escherichia coli

Brief description of procedure: *Escherichia coli* is suspended in two tubes of calcium chloride solution. Into one of the *E. coli* suspensions add the pVIB plasmid. Heat shock the organisms in both tubes. Add suspensions to agar containing the antibiotic ampicillin and incubate.

Discussion: The ampicillin plate with the *E. coli* (no plasmid) should exhibit no growth or have any visible colonies, since *E. coli* is susceptible to this antibiotic. The ampicillin plate with the *E. coli* (with plasmid pVIB) should exhibit growth and have visible colonies. These colonies should also fluoresce, since the plasmid gene confers fluorescence to the *E. coli* if taken up and expressed.

Supplemental Lab 1B
Genetics: Transformation: A Class Demonstration[1]

Whatever is worth doing at all is worth doing well.

Philip Dorner Stanhope

Introduction: Transfer of DNA into organisms has resulted in many advances in medicine, specifically human insulin and human growth hormones. It has also resulted in organisms becoming resistant to antibiotics. It is generally believed that organisms exchange DNA by one of three methods: (1) *transformation*, (2) *transduction*, or (3) *conjugation*. **Transformation**, the incorporation of naked DNA into competent bacteria, was first demonstrated by Frederick Griffith in 1928. He was able to demonstrate that solutions from killed pathogenic strains of *Streptococcus pneumoniae* would make otherwise non-pathogenic *Streptococcus pneumoniae* virulent or pathogenic. Griffith did not know what this transforming agent was, but 16 years later Colin MacLeod, Oswald Avery, and MaClyn McCarty at the Rockefeller, demonstrated that this agent was DNA. **Natural transformation** can occur in a few organisms (Institute for Medical Research), e.g., *Streptococcus pneumoniae, Neisseria gonorrhea, Haemophilus influenza,* and are not as rare as once thought, with virulence being passed between organisms. The DNA must first be transported into the recipient cell and then similarities have to exist on the recipient's DNA for the new DNA to be incorporated (recombination). **Artificial transformation** is accomplished by making recipient cells competent or making the cell membrane permeable to the foreign DNA.

Demonstration of Artificial Transformation
   A. **Materials:**
   1. Brain heart infusion agar (BHIA) stock plate with two organisms streaked on: *Acinetobacter calcoaceticus Str^s* (streptomycin sensitive) and *Acinetobacter calcoaceticus Str^r* (streptomycin resistant)
   2. BHIA plates without streptomycin
   3. BHIA plates with streptomycin
   4. SDS - Sterile 0.5% sodium dodecyl sulfate (0.5ml in screw capped tubes)

   B. **Procedure:** *Instructor will perform these steps*
   1. From the BHIA stock plate, inoculate a small inoculum of *Acinetobacter calcoaceticus Str^r* (streptomycin resistant) in the SDS tube and label it SDS/Str^r. Place this solution in a 60-65^0 C water bath for 30 minutes. Allow to cool to room temperature. This tube will now contain the free (naked) DNA from the Str^r strain and will now be referred to as "DNA."
   2. Label a BHIA plate with four quadrants as follows: Str^r, Str^s, Str^s + DNA, and "DNA."
   3. Inoculate DNA solution onto the DNA marked quadrant, spreading it out to an area the size of a dime.
   4. Repeat this same step on the quadrant marked Str^s + DNA.
   5. Inoculate the quadrant marked Str^s with stock culture and spreading out as per #3.
   6. Inoculate quadrant marked Str^r with stock culture and spread it out as per #3.
   7. Inoculate Str^s cells directly on top of the DNA already inoculated and labeled as Str^s + DNA quadrant.

   C. **Procedure:** *Students will perform this step*

Streak these cultures on a streptomycin containing BHIA plate, divided into three sections and labeled as follows: Str^s, Str^s + DNA, and Str^r , incubate and we will read and interpret the next class period.

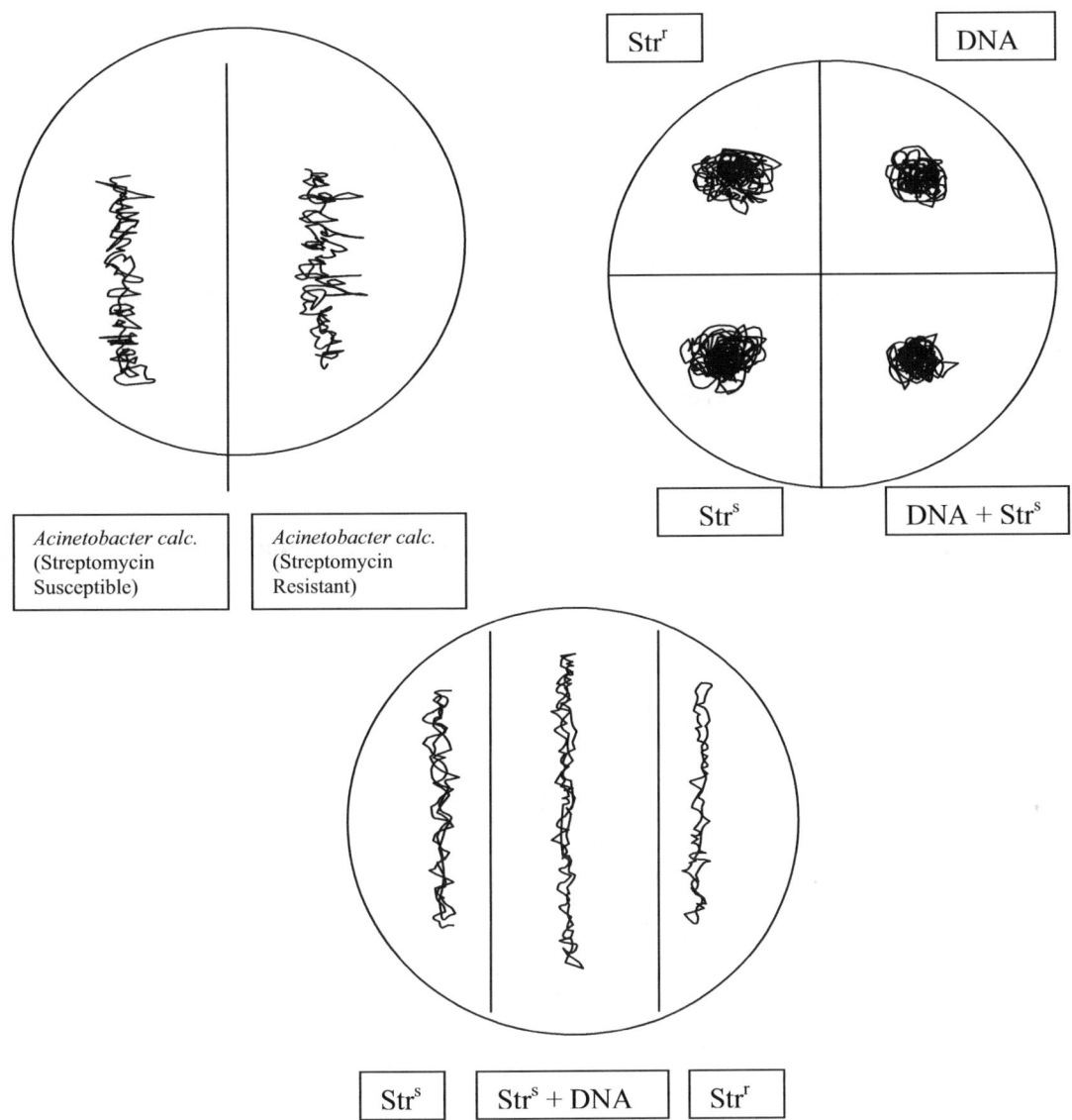

D. **Discussion**

Observe the three quadrants for growth. Determine which grew the most organisms and see if you are able to determine if transformation occurred.

[1]*Transformations: a Teacher's Manual.* (2004). Burlington, NC: Carolina Biological Supply Company.

Supplemental Lab 1C
Clinical Lab: Cavity Prone[1]

Every tooth in a man's head is more valuable than a diamond.

Miguel de Cervantes

Objective: To determine your susceptibility to dental caries

Procedure:

1. Place Snyder test tube agars in a waterbath to liquefy and allow to cool to the touch, approximately 50^0 C.
2. Remove bacteria from teeth by chewing a sterile paraffin block for about three minutes.
3. Deposit the accumulated saliva into a sterile petri dish and discard the paraffin block.
4. With a sterile pipet withdraw 0.25ml of saliva and deposit it in a melted 50^0 C Snyder test tube agar.
5. Rotate the Snyder agar between your palms to mix the saliva and agar
6. Label the Snyder agar and incubate at 35-37^0 C.
7. Check tube at 24, 48, 72, and 96 hours after incubation. Observe the color of the agar to see if the bromcresol green indicator has turned yellow. Record the data on the sheet.
8. After 96 hours, refer to cavity susceptibility table below to determine your susceptibility to dental caries.

0.25 ml

STUDENT DATA SHEET

Date	Day	Hours from Incubation	Time	Color of Snyder Agar
	1	0 hours		Green - control
	2	24 hours		
	3	48 hours		
	4	72 hours		
	5	96 hours		

CAVITY SUSCEPTIBILITY

Interpretation	24 hours	48 hours	72 hours	96 hours
Marked	+	+	+	+
Moderate	-	+	+	+
Slight	-	-	+	+
Negative	-	-	-	+
Negative	-	-	-	-

Supplemental Lab 1D
Glo Germ Hand Washing Experiment

We've known for more than 100 years that hand washing prevents infection, but we still can't get people to wash their hands. Hand washing is the simplest, most effective way to keep from getting sick and making others sick.

Ann Falsey

Purpose: A graphic experiment demonstrating the importance of PROPER hand washing, using a germ simulating gel and a UV light.

Procedure:

1.   Spread germ simulating gel[1] evenly on both your hands, making sure you rub in and around your fingernails. Allow your hands to dry completely (approximately 2 minutes). Place your hands under the UV light.
2.   Subjectively record the "dirtiness," or level of germs on your hands as the amount of fluorescence and record at 0 time as ++++, +++, ++, or + and use a minus sign (-) to indicate clean.
3.   Wash hands for 5 seconds, rinse and check cleanliness with UV light. Record on chart at 5 seconds.
4.   Wash again for an additional 5 seconds and check for cleanliness with UV light. Record on chart at 10 seconds.
5.   Repeat the procedure two more times, for a total of 15 and 20 seconds respectively. Each time record the level of cleanliness.

Results:

Hand Soap	0 Time (control)	5 Seconds	10 Seconds	15 Seconds	20 Seconds

Level of germs prior to washing

6.   Plot a graph with cleanliness on the ordinate and hand washing time on the abscissa.

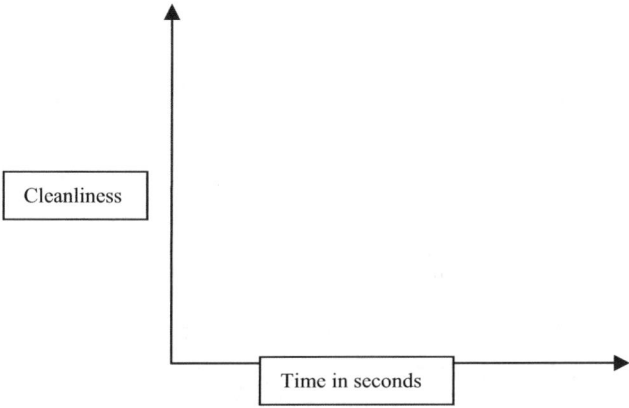

[1] Glo Germ Products. (n.d.). Retrieved March 15, 2010, from http://glogerm.com

Supplemental Lab 1E
Hand Washing Experiment

Proper handwashing with soap and water is an effective barrier to many infectious diseases and promotes better health and well being... Handwashing is one of the most practical and effective ways of preventing the spread of disease.

World Health Organization

Purpose:
To determine the effectiveness of hand washing using soaps with germicidal properties as compared to plain soaps which utilize only mechanical means for sanitization.

Materials:
1. Germicidal hand soaps
2. Mechanical hand soap without bactericidal or bacteriostatic properties
3. Waterless hand sanitizers
4. Tryptic soy agar plate

Procedure:
Mark the bottom of a petri dish with 4 quadrants. Label quadrant #1 Control A, label quadrant #2 Mechanical Hand Soap A, label quadrant #3 Control B, and label quadrant #4 Germicidal Hand Soap B.

Lab partner A touches quadrant #1 with two fingers of the right hand. Partner A then washes hands with mechanical hand soap as he/she sings "Happy Birthday" two times *silently* (15-20 seconds)—rinses well and then dries with a clean paper towel. Partner A then touches quadrant #2 with the same two fingers as previously done with the control.

Lab partner B touches quadrant #3 with two fingers of the right hand. Partner B then washes hands with Germicidal Hand Soap as he/she sings "Happy Birthday" two times *silently* (15-20 seconds)— rinses well and then dries with a clean paper towel. Partner B then touches quadrant #4 with the same two fingers as previously done with the control. Incubate at 35-37^0 C for 24-48 hours.

Results:
Estimate the contamination on the plate by assuming the quadrants prior to washing are the controls and label as +++++. Quadrants after washing label as either same as controls +++++, or estimate reduction of contamination and label as; ++++, +++, ++, +, or – (no contamination). Record your results in the table below.

Subjects	Soap	Control (Prior to washing hands)	Result (After washing hands)
Partner #1-		+++++	
Partner #2-		+++++	

Discussion:
1. *Was there a difference between the antimicrobial soaps and the regular soap?*

2. *After performing this very simple experiment what do you feel are the parameters affecting reduction of microbial loads after washing?*

3. *In your opinion, do you feel antimicrobial soaps should be used for every hand washing?*

Supplemental Lab 1F
Fermentation Lab
Yogurt Lab (Fermentation of Milk)

Yoghurt is very good for the stomach, the lumbar regions, appendicitis and apotheosis.

Romanian Writer 1912

Introduction:
Yogurt can be made with any milk, with the flavor and texture dependent on the kind of milk chosen. The richer the milk, i.e. the more butterfat it contains, the less tart and the thicker the yogurt will be. Goat's milk contains shorter protein molecules than cow's milk and can be tolerated by individuals sensitive to milk; however, more delicate procedures must be employed in the manufacturing steps of goat yogurt, specifically lower temperatures of heat. Yogurt is a fermented product which must contain both Lactobacillus bulgaricus and Streptococcus thermophilus and have an initial fermentation count of 10^8 cfu/g of the cultures to display the National Yogurt Association seal of approval.

Enzymes from the bacteria (lactase) convert the sugar in the milk (lactose) into lactic acid. The lactic acid changes the conformation of the proteins in the milk and thus the milk becomes viscous along with the accompanying sour taste (lactic acid).

note: lactose intolerant individuals lack the enzyme lactase and thus cannot digest milk

Procedure:
1. Add 150ml of whole milk to a 250ml clean beaker.
2. Add 15g of powdered milk to the milk to make a thicker yogurt.
3. Heat with constant stirring to approximately 88^0 C.
4. Remove from heat source and allow cooling to 44^0 C.
5. Add 1 teaspoon of starter culture and stir gently but mix well.
6. Cover and incubate for 24 hours at 37-44^0 C, preferably 44^0 C.
7. After incubation, remove and refrigerate.
8. Ready to add fruit, flavors, etc. and will taste next class and critique.

Cultures: Standard Yogurt vs. Probiotic Yogurt

Yogurt #1 *Streptococcus thermophilus, Lactobacillus bulgaricus* are the standard yogurt cultures specified by The National Yogurt Association (NYA)[1].

Yogurt #2 *Lactobacillus acidophilus, Streptococcus thermophilus, Lactobacillus bulgaricus,, Bifidobacter spp., Lactobacillus casei, Lactobacillus reuterii, Lactobacillus rhamnosus* are found in probiotic yogurt. Note: may contain all of the above cultures or just some of them.

FERMENTABLE SUGAR: Lactose – a disaccharide

Name: _____

Date: _____

Starter Culture (#1, #2,): _____

Additives (fruit, granola, etc): _____

Taste Analysis:_____   Consistency: _____
  1=excellent, 2=good, 3=OK, 4=fair, 5=poor, 6=unfit for human consumption

[1]National Yogurt Association (NYA). http://aboutyogurt.com/index.asp?bid=5

Supplemental Lab 1G
Fermentation Lab
Sauerkraut Laboratory (Natural Fermentation of Cabbage)

When General Lee took possession of Chambersburg on his way to Gettysburg, among the first things he demanded for his army was twenty-five barrels of Saur-Kraut.

The Guardian 1869

Introduction: Sauerkraut is a naturally fermented cabbage. Natural fermentation is one of the oldest means of food preservation. Natural fermentation of the cabbage is a result of the bacteria indigenous to the cabbage, producing lactic acid and some minor by-products, in the presence of 2-3% salt. Salting of the cabbage serves two major purposes. First, it causes an osmotic imbalance which results in the release of water and nutrients from the cabbage leaves. The fluid expelled is rich in sugar and growth factors. Second, the salt concentration inhibits the growth of many spoilage organisms and pathogens, but it does not inhibit the desired floral succession. Since this is a wild or natural fermentation no starter cultures are added and thus the fermentation relies upon the bacteria normally present on the cabbage. The floral succession is governed by the pH of the growth medium and in general proceeds in the manner listed below.

1. Coliforms (a general class of gram negative lactose fermenting organisms found in the intestinal tract of warm blooded animals) start the fermentation, with *Klebsiella pneumoniae, Klebsiella oxytoca,* and *Enterobacter cloacae* as the dominant organisms.
2. After some acid is produced (pH drops), *Leuconostoc spp.*, a heterofermentive lactic acid bacteria grows and produces more acid (pH drops even further).
3. The final stage has *Lactobacillus spp.* and occasionally *Pediococcus spp* now as the dominant growth, producing lactic acid and some minor by-products. This provides the characteristic taste of sauerkraut.

Procedure:

1. The cutters and slicers, *schneidens* and *abschneidens*, trim the cabbage by removing the outer leaves. Save these outer leaves, since they will be used as the sacrificial leaves for the top of the container. Wash the trimmed cabbage heads with warm water. Cut in half and remove the hard central core. Then cut in half, so as to have four quarter sections for each head. Deliver to the shredders.
2. The shredders, *zerfetzens*, now shred the cabbage with the slaw cutter. Note: since we do not want sausages in the cabbage at this stage, please follow instructor's directions and be very careful. Deliver to the weighers.
3. The cabbage weighers, kohl *abwiegeners*, weigh out the shredded cabbage in 5lb increments or 2,268g and add them to the fermentation vessel. The salt weighers, salz abwiegeners, weigh out 3 tablespoons of salt or 42.51g for each 5lb layer of cabbage. Mix by hand.
4. The stompers, *stampfeners*, add a little water to the first layer of cabbage and salt and then begin stomping with the wooden stomper until juicy. Continue the process of adding cabbage, salt, and stomping until the level reaches approximately 8-12 inches from top of vessel.

5. The finishers, *absolvierens*, add the clean outer sacrificial leaves to the top of the stomped cabbage/salt mixture. A sacred plate is then placed on top of the sacrificial leaves and then submerged. A sacred rock is placed on the sacred plate to maintain anaerobic conditions. Fermentation is allowed to proceed for 5-6 weeks at room temperature, after which bagging the sauerkraut takes place, followed by qualitative testing (or in our case the eating of the kraut).

FERMENTATION

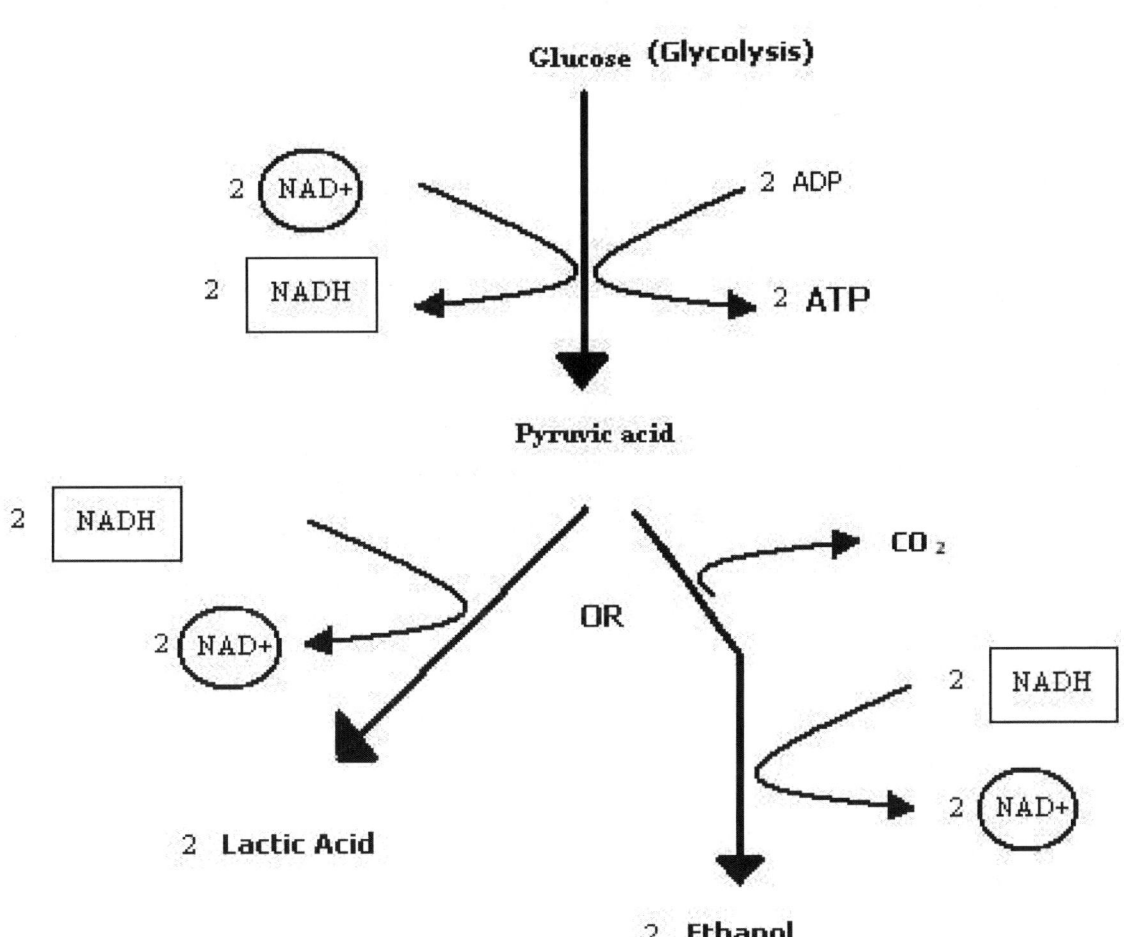

Supplemental Lab 1H
Fermentation Lab
Root Beer/Birch Beer (Fermentation of Roots)

There is more similarity in the marketing challenge of selling a precious painting by Degas and a frosted mug of root beer than you ever thought possible.

A. Alfred Taubman

Introduction:

Root and birch beers are commonly found in markets throughout Lancaster County, primarily in the summer months. The first documented evidence of the production of these fermented products was in the American Colonies. Dr. Chase's 1869 recipe had you take hops, burdock, yellow dock, sarsaparilla, dandelion, and spikened roots and boil, then strain and add sugar, molasses, and yeast. Allow to ferment for a couple of hours and bottle. In 1960, the FDA outlawed sassafras, one of the spikened roots, because it contained safrole, a proven carcinogen in lab rats.[1]

We will dispense with the tedium of boiling spikened roots and use an extract of either root beer or birch beer. Shank's Extracts of Lancaster, Pennsylvania is a wonderful resource for these extracts. These aromatic roots provide the taste, but the sugar provides the nutrients for the fermentation process.

Procedure:

1. Dissolve 1809g of sugar in 17980ml of warm water.
2. Stir in one bottle of either Shank's root beer or birch beer extract,[2] 118ml.
3. Add 10ml of prepared ale yeast, *Saccharomyces cerevisiae*, a 24-hour culture.
4. Stir in yeast and bottle immediately in pre-cleaned bottles, 2.5cm from the top.
5. Let stand in a warm place (75°–80° C) for 10 days.
6. Refrigerate before serving.

Culture: *Saccharomyces cerevisiae*

Name: _____   **Date:** _____

Root Beer/Birch Beer: _____

Taste Analysis: _____   Effervescence: _____
1=excellent, 2=good, 3=OK, 4=fair, 5=poor, 6=unfit for human consumption

---

[1]Root Beer. (n.d.). Retrieved August 13, 2010, from http://www.root-beer.org/
[2]Shank's Extracts Inc. Flavoring Extracts and Syrups. 350 Richardson Drive, Lancaster, PA, 17603-4034. 717-393-4441.

Supplemental Lab 1I
Fermentation Lab
Kombucha Laboratory (Fermentation of Sweet Tea)

"Kombucha Mushroom People Sitting Around all Day"

Lyrics from the song "Sugar" by *System of a Down*

Introduction: **Kombucha** is a fermented product of sweetened black tea, which has been inoculated with what is affectionately termed a kombucha mushroom. This mushroom is actually a symbiotic mixture of numerous acetic acid producing bacteria and several yeast organisms held together by a membranous substance. The longer the sweetened tea is allowed to contact the mushroom, the more acidic the tea becomes, and eventually the tea has the distinct taste of vinegar. After 24 hours the pH decreases to 1.8, which according to CDC should prevent the survival of most potentially contaminating organisms.[1]

Procedure:

1. Add 100g of sugar to 1000ml of pure water and boil for 1-3 minutes.
2. Allow to cool and then add 3-4 tea bags of desired tea.
3. To this sugar tea mixture add about 50 to 100ml of previously fermented kombucha tea (last batch) and then add the mushroom. Note: the baby mushroom will form on top of the mother mushroom and either one may be used for the next fermentation.
4. Allow to ferment at room temperature for 7-20 days, depending on the sweetness desired of the final kombucha tea.
5. After removing the mushroom and 50 to 100ml of the fermented tea, filter the kombucha through a cloth filter to further remove most of the sediment and store under refrigerated conditions.

[1] Centers for Disease Control. (1995). *Unexplained Severe Illness Possibly Associated with Consumption on Kombucha Tea, Iowa 1995.* Morbidity and Mortality Weekly Report, 44(48), 892-893. Retrieved March 12, 2010 from www.cdc.gov/mmwr/preview/mmwrhtml/00039742.htm

Special Thanks to the Chuprin family of New Holland, PA for supplying the original Kombucha mushroom.

68

Supplemental Lab 1J
Heterotrophic Plate Count

Theory guides. Experiment decides.

Ann Folsey

Introduction: This lab involves the counting of visible microbial colonies using a Quebec Colony Counter, which has a magnifying lens and a lighted surface incorporating a Wolffheugel grid. Each colony, i.e., colony forming unit, represents the progeny of either one cell or a group of cells, which have generated enough organisms, to be visible to the naked eye.

Procedure:

1. For greatest accuracy count those plates having 25-250 cfu's/plate. Count colonies on the plate by following a back and forth pattern.

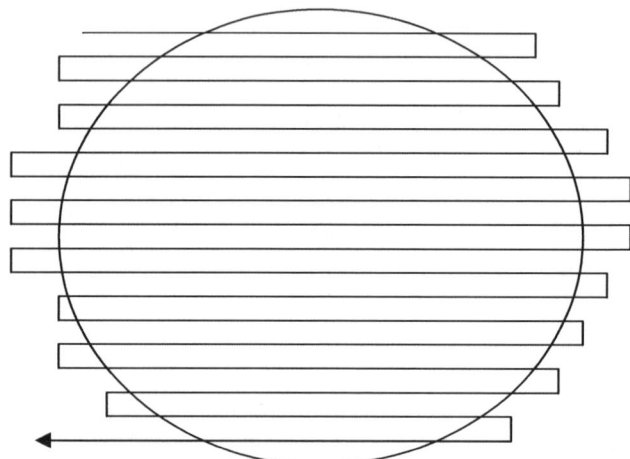

2. If there are no colonies on the plate report as <1 cfu as opposed to 0 cfu.

3. If there are more than 250 cfu's/plate do not report as TNTC/plate, but report as follows:

   A. If there are <10cfu's/1sq cm (1 large square on grid), count the number of cfu's on 13 squares (7 squares across and 6 squares down) add together and multiply by 4.38. This is the estimated cfu's/plate.

   B. If there are >10 cfu's/1sq cm, count any 4 squares, divide by 4 and multiply by 57. This is the estimated cfu's/plate.[1]

4. If you are reporting number of colonies as 3 digits, round off the last digit, e.g., report 142 as 140. If reporting as 2 digits, do not round off, e.g., report 14 as 14.

5. Colonies which are imbedded in the agar (subsurface colonies) will be much smaller than surface colonies. Spreaders are colonies that grow in such a way that they appear to be spread across the plate. Count the following types of spreaders as 1 colony forming unit or 1cfu: a chain of bacteria caused by disintegration of a bacterial clump, a film of growth between the agar and bottom of the plate, and a colony forming in a film of water at the edge of a plate or over the surface of the plate.

[1]Hach Company. (2000). *Methods 8241 & 8242: For Water and Wastewater.* Retrieved July 12, 2010, from http://www.water-research.net/Waterlibrary/watermanual/platcount.pdf

Supplemental Lab 1K
Demonstration of Transmission of Microbes using Glo-Germ Powder

The ideal way to get rid of any infectious disease would be to shoot instantly every person who comes down with it.

H.L. Mencken

Introduction: Microbes can easily be transmitted from inanimate objects, specifically known as *fomites*, to susceptible individuals. Examples of some of these articles of transmission are neckties, lab coats, stethescopes, urinals, bed linens and a whole list of other sundry items. One glaring example was the anthrax attack on this country perpetrated by Bruce Ivins, an army biologist, working out of Fort Detrick, Maryland. Bruce Ivins single handedly mailed anthrax-laden letters to New York news media offices and Washington senate offices, which resulted in 5 deaths and illness in 11 others.[1] This is a rather simple demonstration to determine just how microorganisms can be transmitted via a *fomite* route.

Procedure: Papers to be handed out to the students are brushed with Glo-Germ powder,[2] a fluorescent generating, non-toxic powder. These papers are stacked in a pile on the instructor's desk and the students are asked to take a paper and return to their lab station. The information on the handouts is then discussed, allowing the students ample time to handle the powder infested paper. At a convenient time in the class schedule for that day, but prior to any student's washing their hands, a UV light is passed around the class, so that students may view their hands to see where fluorescence has occurred. *Note: turning the room lights off makes for a more dramatic exhibition of the fluorescence.* If enough time has passed, some of the students may have touched their faces and/or mouth and nose, whereby fluorescence will be displayed around these areas. My experience has been to coincide this study with the completion and interpretation of the environmental lab, where tabulated data from all the labs can then be analyzed. This allows for a longer contact time for the students with the Glo-Germ brushed sheets and more students tend to test positive.

[1] Warrick, J. (2010). "FBI Investigation of 2001 Anthrax Attacks Concluded; US Releases Details". *Washington Post.* http://www.washingtonpost.com/wp-dyn/content/article/2010/02/19/AR2010021902369.html

[2] Glo Germ Products. (n.d.). Retrieved March 15, 2010, from http://www.glogerm.com/

Supplemental Lab 1L
Antibiotic Susceptibility: Different Strains of an Organism

The trouble with being a hypochondriac these days is that antibiotics have cured all the good diseases.

Caskie Stinnett

Introduction: Antibiotic resistance has become a major concern in the healthcare system. While growing concern is usually directed at methicillin resistant *Staphylococcus aureus* (MRSA), there are other species of bacteria that show less susceptibility to particular antibiotics when measured among different strains of the same species. Because of these different strains and their different susceptibilities to antibiotics, it is not only important to identify the infectious agents, but to also identify their susceptibility to the antibiotics that will be chosen to combat the infection.

Procedure:

Students will have copies of Mueller-Hinton, antibiotic susceptible plates, on laminated cards, inoculated with different strains of one species of bacteria, along with different antibiotic discs.

Organisms:
Escherichia coli
Serratia marcescens
Pseudomonas aeruginosa

Brief description of procedure: From the laminated cards provided,[1] students will measure the zone of inhibition of each antibiotic with a metric ruler (mm). The students will chart their measurements for each strain and compare susceptibility of each strain with the corresponding antibiotic.

Discussion: This laboratory is designed to show that some of the more popular antibiotics are becoming less useful across the board as the bacteria are evolving defenses against them and becoming increasingly more resistant. This demonstration shows both the need for new antibiotics as well as the increasing costs necessary for their development.

---

[1] Reproduced with permission of the ASM MicrobeLibrary (http://www.microbelibrary.org)

Table for Supplemental Lab 1L

Serratia marcescens

Pseudomonas aeruginosa

Escherichia coli

Interpretation of Inhibition Zones of Test Cultures[1]

Antimicrobial Agent	Disk Content	Resistant mm or less	Intermediate mm range	Susceptible mm or more
Kanamycin	30mcg	13	14-17	18
Novobiocin	30mcg	17	18-21	22
Penicillin	10iu	28	-------	29
Streptomycin	10mcg	11	12-14	15
Tetracycline	30mcg	14	15-18	19

Gram-positive discs
Novobiocin (N)
Tetracycline (T)
Streptomycin (S)
Penicillin (P)

Gram-negative discs
Kanamycin (K)
Chloramphenicol (C)
Tetracycline (T)

[1]Cappuccino, C. & Sherman. N. (2002). *Microbiology, A Laboratory Manual, 6th edition.* San Francisco, CA: Benjamin Cummings

MICROBIOLOGY
FIELD TRIPS

**Lancaster
Cemetery**

**Lancaster Brewing
Company**

**Lancaster General
Hospital Micro Lab**

Field Trip to Lancaster Cemetery
INFECTIOUS DISEASES
Deaths from U.S. Military Conflicts[1]

Conflict	Hostile Deaths	Nonhostile Deaths	Total Deaths
Persian Gulf 1991	147	235	382
Somalia 1992-94	30	14	44
Vietnam 1964-1973	47,414	10,789	58,203
Korean War 1950-1953	33,741	2,827	36,568
WW II 1941-1946	291,557	113,842	405,399
WW I 1917-1918	53,402	63,114	116,516
*Civil War 1861-65			Ca. 700,000
Total			**1,317,112**

*Estimates since approximately 80% died from complications from wounds and communicable diseases

Timeline of Infectious Diseases in Eastern Pennsylvania[2]

Date	Location	Disease	Date	Location	Disease	Date	Location	Disease
1788	Phila	Measles	1793	Harrisburg	Yellow Fever	1794	Phila	Yellow Fever
1796-98	Phila	Yellow Fever	1820	Schuylkill River	Fever	1837	Phila	Typhus
1860	All Counties	Smallpox	1885	Plymouth	Typhoid	**1918**	**All Counties**	**Spanish Flu 1918**

*Most serious epidemic of late 18th and 19th century--20 million deaths worldwide

Epidemics of the Old World[3]

Date	Area	Outcome
430 BC	Athens, Greece	Unidentified--200,000 infected 65,000 dead
160 AD	China-collapse of Han Empire	Bubonic plague
166 AD	Antonine plague of Rome (led to collapse of Roman Empire)	4-7 million Europeans killed Combination smallpox, measles, bubonic plague
1346-1350	China to Russia to Europe	Bubonic plague--killed 1/3 population of Europe
1492	Caribbean to Americas to Europe	Influenza, smallpox, tuberculosis, gonorrhea Killed 8 million where Columbus set foot
1542	Egypt to Europe	Bubonic plague--killed 40% population of Constantinople
Early trading period 1500's	Africa	Malaria, yellow fever, dysentery (probably cholera)--termed the "White Man's Grave"
16th century	Europe-Africa-Central & South Americas	Tuberculosis & viral diseases-decreased population of Mexico by 95% in 75 years
16th to 19th century	Europe & Pacific Islands	Tuberculosis & viral diseases-decreased population of Pacific Islands by 95%

Major U.S. Epidemics[1]

Date	Area	Disease-Deaths	Date	Area	Disease-Deaths
1793	Philadelphia	Yellow Fever-4000	1878	Southern States	Yellow Fever-13000
1832	New York City	Cholera-3000	1916	Nationwide	Polio-7000
1832	New Orleans	Cholera-4340	1918	Nationwide	Flu-500,000
1848	New York City	Cholera-5000	1949	Nationwide	Polio-2720
1853	New Orleans	Yellow Fever-7790	1952	Nationwide	Polio-3300
1867	New Orleans	Yellow Fever-3093	1981-	Nationwide	AIDS-457,667

[1]Leland, Anne. American War & Military Operations: Lists & Casualties. www.fas.org/sgp/crs/natsec/RL32492.pdf
[2]www.redboneheritagefoundation.com/Chronicles/Disease%20Epidemics%20in%20Early%20America.htm
[3]Staley, J., Gunsalus, R., Lory, S., Perry, J. (2007). *Microbial Life 2nd Edition.* Sunderland, MA: Sauter Associates. PP 32-33.

Field Trip to Lancaster Cemetery

1918 Spanish Influenza Pandemic and Lancaster, Pennsylvania[1]

Introduction

1. Caused more deaths in the United States then World War I--500,000 deaths.
2. Originated in the U.S.--March 1918 rather than the far east as other influenza outbreaks (pigs, ducks, and humans in close proximity in the Far East).
3. Americans carried disease to Europe during WW I.
4. Afflicted primarily people in the prime of life, 29-40 year olds.
5. Pneumonia and renal failure complications following the flu.
6. New York City was the least affected of the major U.S. cities.
7. Frank Milley, an American soldier, was the first to die of flu.
8. Spanish influenza was the name falsely attributed to the epidemic.
9. Cholera was the epidemic responsible for the most deaths up until the Spanish flu.

Lancaster City

1. 3000 cases as of Oct. 6, 1918--Lancaster had a population of 52,500.
2. 301 dead as of Nov. 5, 1918; however, was an underestimate, actual deaths were more realistically 361.

Lancaster City Conditions in 1918--Fear of the Epidemic

1. Oct. 4, 1918 Lancaster Board of Health (LBH) advised closing all public places of gathering, prohibited all public events, prohibited visiting friends unless critically ill, made funerals private, churches and schools closed by discretion, people encouraged to walk to work, streetcars mandated to have all windows opened when in service, all city streets ordered to be cleaned, Central Market ordered to be washed and fumigated, quarantine cards ordered to be placed on all homes of victims and left on until three days after the last person recovered.
2. Franklin & Marshall closed, city schools closed, one half of the county schools closed.
3. Oct 4, 1918--23 patients at Lancaster General Hospital had serious complications from the flu, 30 patients at St. Joseph's Hospital with serious complications from the flu.
4. Moose and Elk's Clubs of Lancaster, opened and staffed as emergency hospitals.
5. All churches closed for mass worship, except St. Mary's Catholic church, which secretly kept a side door opened for parishioners (missing mass a mortal sin)--Oct. 10, 1918 Health Officer of Lancaster secured the door.
6. Lancaster was only hit with one wave of the flu, contrary to the rest of the country and the rest of the world which had two waves.

[1] From *Journal of the Lancaster County Historical Society, Vol. 102, No. 4 (Winter 2000)* by Meg R. Gerstenblith. Copyright © 2000 by Lancaster County Historical Society. Reprinted by permission.

Lancaster Cemetery - 1st Reformed Church (United Church of Christ)

After main gate on Lemon--20 feet and on left
1. J.F. Lutz 1885-1918--age 33 died of Spanish Influenza.
2. Many gravesites at Greenwood Cemetery in the city have records to indicate cause of death and many influenza victims interred there--records are at Cemetery and on microfilm at Lancaster Historical Society.

After main gate on Lemon--20-30 feet on left large area for Hartman Family
1. John Hartman 1831-1899, age 68, father
2. Charles Crier Hartman 1866-1867, 1 year
3. Elizabeth Messersmith 1879-1880, 1 year
4. Florence Hoffmeier 1874-1874, <1 year
5. Clarence Barker 1876-1881, 5 years
6. Ralph Bacon 1886-1886, <1 year
7. Granville Hartman 1878-1878, <1 year

Left after Hartman Family
1. Marge Kauffman, age 81, Edward Kauffman, age 85, parents
2. John Kauffman 1837-1839, age 2
3. Mary Kauffman 1829-1837, age 8

Left after Hartman Family
1. Thomas Apple, age 69, Emma Apple, age 94, parents
2. Maude Apple 1867-1880, age 12
3. Elizabeth Apple 1854-1876, age 21

Augusta Harriet
Daughter of Charles M. and Emelia Bitner
1884-1906
Could Love Have Kept Her?

The inscription immortalizes the brief life of this young woman and her tragic death. Accounts vary as to the exact time, place, and manner of her death; however, the following is a documented account taken from Dorothy Fiedel's *Official Ghost Guide to Lancaster County:*[1]

Augusta, who was from a fairly wealthy family, fell in love with a young man named Tevis. Her parents did not approve of this affair, but she defied them and went ahead with her wedding plans. On the day of her wedding she had one final argument with her parents and in anger vowed to leave and never see her parents again. She turned to run out the front door, but as she reached the top of the stairs, she fell and died of a broken neck. On the anniversary of her death, it is said that Augusta's spirit roams the cemetery in search of her beloved Tevis, but Tevis does not rest in this cemetery.

Descendents of her family have tried to stop this legend, saying she died of Tuberculosis and this is also the cause of death listed on the cemetery record (note: high fevers are common with Tuberculosis along with weakness and dizziness).

Ghost hunters have recorded light orbs around the monument, indicative of spirits.

[1]Fiedel, D. (2002). *Fiedel's Official Ghost Guide to Lancaster County, Pennsylvania*. Lancaster, PA: Fisher Productions.

Field Trip to Lancaster Brewing Company
Demonstration of Fermentation

Grains (barley, wheat, rye, etc.) are allowed to germinate to release sugars into what is termed malt. The malt is solubilized in water and boiled and this sugar solution is called wort. Hops, the female flower cones of the hop plant, are added to impart the bitter tangy flavor to the beer. The sugars in the wort then undergo a series of chemical reactions to produce alcohol, residual organics, and carbon dioxide.

$$C_6H_{12}O_6 \underline{\text{ Glycolytic Enzymes }} 2C_3H_4O_3 \underline{\text{ Decarboxylation }} 2C_2H_4O \underline{\text{ Reduction }} 2C_2H_5OH + CO_2$$

Glucose   **Pyruvic acid**   **Acetaldehyde**   **Ethyl alcohol**

Fermentations by Naturally Occurring Organisms

Product	Application	Organism
Beer, Wine, & Spirits	Alcohol production	*Saccharomyces spp.*
Bacitracin	Antibiotic	*Bacillus subtilis*
Chloramphenicol	Antibiotic	*Streptomyces venezuelae*
Citric acid	Food flavoring, medicine	*Aspergillus niger*
Erythromycin	Antibiotic	*Streptomyces erythaeus*
Invertase	Candy	*Saccharomyces cerevisiae*
Lactase	Digestive aid	*Escherichia coli*
Neomycin	Antibiotic	*Streptomyces fradiae*
Pectinase	Fruit juice	*Aspergillus niger*
Penicillin	Antibiotic	*Penicillium notatum*
Riboflavin	Vitamin	*Ashbya gossypii*
Streptomycin	Antibiotic	*Streptomyces griseus*
Subtilisins	Laundry detergent	*Bacillus subtilis*
Tetracycline	Antibiotic	*Streptomyces aureofaciens*

Fermentations by Genetically Engineered Organisms

Product	Application	Organism
Bovine growth hormone	Milk production cows	*E. coli*
Cellulase	Cellulose	*E. coli*
Human growth hormone	Growth deficiencies	*E. coli*
Human insulin	Diabetics	*E. coli*
Monoclonal antibodies	Therapeutics	Mammalian cell culture
Ice-minus	Prevents ice on plants	*Pseudomonas syringae*
Sno-max	Makes snow	*Pseudomonas syringae*
t-PA	Blood clots	Mammalian cell culture
Tumor necrosis factor	Dissolves tumor cells	*E. coli*

Microbiology - An Anthology

DeKruif, P. (1926). *Microbe Hunters.* New York,NY: Harcourt Brace and Co.

Those who cannot remember the past are condemned to repeat it. Those who dwell in it have no future, so remember the past but embrace the future.

George Santayana

GOLDEN AGE OF MICROBIOLOGY (1857-1907)

Antoni VanLeeuwenhoek - Father of Microbiology (1635-1723)

Highly Refined Lens Grinder

Discovery of Microbes or Animalcules

Spontaneous Generation - Abiogenesis vs. Biogenesis

Aristotle 384 BC- Spontaneous Generation

Francis Redi- Decaying Meat Experiment

John Needham - Corked vs. Uncorked Jar Experiment

Lazzaro Spallanzani - Sealed vs Unsealed Flasks Experiment

Louis Pasteur - Swan Neck Flask Experiment

Louis Pasteur-Chemist by trade

Fermentation and Spoilage

Germ Theory of Disease

Vaccines

Robert Koch-Medical Doctor by trade

Discovered Causative Agents of Cholera and Anthrax

Koch's Postulates

Accomplishments in Microbiology

Ignaz Semmelweis

Joseph Lister

Florence Nightingale

Paul Ehrlich

Alexander Fleming

Chapter #1 Introduction

Bauman, Robert. (2009). *Microbiology With Diseases by Taxonomy, 3rd Edition*. San Francisco, CA: Benjamin Cummings.

Science is wonderfully equipped to answer the question "How?" but it gets terribly confused when you ask the question "Why?"

Erwin Chargaff

Microbes/Microorganisms

Bacteria, Fungi (Yeasts & Molds), Protozoa, Algae, Viruses

>87% Beneficial

10% Opportunistic Pathogenic

<3% Pathogenic

Types of Microorganisms

Bacteria (prokaryotes)

Archaea (prokaryotes)

Fungi (eukaryotes)

Protozoa (eukaryotes)

Algae (eukaryotes)

Viruses (infectious particles/non-living)

Modern Age of Microbiology

Chapter #4 Microscopy, Staining, & Classification

Bauman, Robert. (2009). *Microbiology With Diseases by Taxonomy, 3rd Edition*. San Francisco, CA:Benjamin Cummings.

Where the telescope ends, the microscope begins, which of the two have the grander view?

Victor Hugo

Units of Measurement

Metric System - unit of length = Meter

Conversions

Size of Microbes

Viruses

Bacteria

Fungi

Yeasts

Molds

Protozoa

Microscopes

Resolution (*R = 0.61 X Wavelength/ numerical aperture*)

Simple vs. Compound

Dark Field

Phase Contrast

Fluorescent

Electron Microscopes

Transmission (TEM)

Scanning (SEM)

Staining

Simple Stains

Differential Stains

Structural Stains

Classification of Microorganisms

Carolus Linnaeus

Robert Whittaker – 5 Kingdoms

Carl Woese – 3 Domain System

Chapter # 3 Cell Structure & Function

Bauman, Robert. (2009). *Microbiology With Diseases by Taxonomy, 3rd Edition*. San Francisco, CA:Benjamin Cummings.

Man presumes he is a fallen angel, when in actuality he is a risen ape

Desmond Morris *The Naked Ape*

EUKARYOTES VS. PROKARYOTES – COMPARISON[1]

PROKARYOTIC CELL	EUKARYOTIC CELL
Cell Wall - Peptidoglycan (Murein)	Cell Wall - some have, some do not (Plants-cellulose, Fungi-chitin)
Cell Coat - Capsule or Slime Layer	No equivalent structures
Cell Membrane - lacks Sterols (e.g. Cholesterol)	Cell Membrane - contain Sterol (e.g. Cholesterol)
Few Membrane - Bound Organelles	Many Membrane Bound Organelles (e.g. Mitochondria, ER, Chloroplasts, Golgi Bodies)
Respiration in Cytoplasm/Cell Membrane Fermentation in Cytoplasm	Respiration in Mitochondria Fermentation in Cytoplasm
Small Ribosomes - 70s (30s + 50s subunits)	Larger Ribosomes - 80s (40s + 60s subunits)
DNA Free in Cytoplasm - Nucleoid Region Genome - Haploid & Naked - 1 chromosome in most	DNA Membrane Bound - Nucleus Genome - Diploid + Histones - >1 Chromosomes
Flagella - Simple - No Cilia	Flagella - Complex - can have Cilia
Endospores (*Bacillus & Clostridium*)	No Endospores - Cysts in Some Protozoa
Pili (gram-negative bacteria)	No equivalent structures

[1]Hairston, Robert. (2001). *Biology 221: Student Study Guide and Laboratory Manual*. Dubuque, IA: Kendall/Hunt

Eukaryotic Cell Structure

Cell Wall (animals & most protozoa lack)

Cell Membrane (fluid mosaic)

Nucleus

Endoplasmic Reticulum

Ribosomes

Golgi Complex

Lysosomes & Peroxisomes

Mitochondria & ATP

Motility

Prokaryotic Cell Surface Structures

Cell Coat = Glycocalyces - smooth vs. rough colonies

Cell Wall

Cell Membrane/Plasma Membrane

Classification of Bacteria Via Cell Wall Structure (gram stain) - *CHAPTER 4*

Gram-Positive Organisms

Gram-Negative Organisms

Prokaryotic Cell Organelles

 Ribosomes

Prokaryotic Genetic Structure

 DNA

 Chromosome

 Plasmids

Prokaryotic Special Structures

Flagella (Rotate 360^0)

 Monotrichous

 Lophotrichous

 Amphitrichous

 Peritrichous

Axial Filaments

Endospores

Pili & Fimbrae

Chapter # 11 – Characterizing & Classifying Prokaryotes

Bauman, Robert. (2009). *Microbiology With Diseases by Taxonomy, 3rd Edition.* San Francisco, CA:Benjamin Cummings.

The greatest poverty in life is the feeling of not being wanted

Mother Theresa

Bacteria - Synonomous with Microbiology

Prokaryotic cells

Cell walls contain peptidoglycan (exception mycoplasmas & archaea)

Size = 1micrometer (approximately)

Classification of bacteria (*Bergey's Manual of Determinative Bacteriology*)

1. **Cell morphology**
2. **Staining characteristics**
3. **Motility**
4. **Colonial morphology**
5. **Atmospheric requirements**
6. **Nutritional requirements**
7. **Biochemical activities**
8. **Pathogenicity**
9. **Amino acid sequence of proteins**
10. **Genetic composition**

Cell Morphology

Rods - Bacillus

Spherical - Cocci

Spiral

Pleomorphism vs. Monomorphism

Arrangements of Prokaryotic Cells

Diplococci

Streptococci

Tetrads

Sarcinae

Staphylococci

Bacilli

Colonial Morphology

Reproduction of Prokaryotic Cells

Endospores

Low G + C Gram-Positive Bacteria

High G + C Gram-Positive Bacteria

Gram-Negative Proteobacteria

    Alphaproteobacteria

    Betaproteobacteria

    Gammaproteobacteria

    Deltaproteobacteria

Other Gram-Negative Bacteria

Archaea

    Prokaryotic Cells

Chapter #12 – Characterizing & Classifying Eukaryotes

Bauman, Robert. (2009). *Microbiology With Diseases by Taxonomy, 3rd Edition*. San Francisco, CA:Benjamin Cummings.

We can allow satellites, planets, suns, universe, nay whole systems of universes, to be governed by laws, but the smallest insect, we wish to be created at once by a special act.

Charles Darwin

General Characteristics of Eukaryotic Microorganisms

Organelles

Reproduction

Asexual

Sexual

Classification of Eukaryotes

Protozoa–Unicellular

Cell walls = none

Locomotion

Protozoa diseases

Dysentery–*Entamoeba histolytica*
Vaginitis–*Trichomonas vaginalis*
Malaria–*Plasmodium spp.*
Sleeping Sickness & Chagas Disease–*Trypanosoma spp.*
Leishmaniasis (Desert Storm)–*Leishmania spp.*
Diarrhea–*Giardia lamblia & Cryptosporidium parvum*
Toxoplasmosis–*Toxoplasma gondii*

Fungi–Unicellular & Multicellular

    Cell walls = chitin

    Reproduction

    Fungal diseases

        Stachybotris
        Tinea
        Histoplasmosis
        Blastomycosis
        Coccidiomycosis
        Cryptococcal meningitis

Algae–Unicellular, Colonial, or Multicellular

    Cell walls = multiple (agar)

    Algae diseases

        Ciguatera
        Paralytic shellfish poisoning
        Amnesiac shellfish poisoning

Chapter #13 – Characterizing & Classifying Acellular Infectious Agents

Bauman, Robert. (2009). *Microbiology With Diseases by Taxonomy, 3rd Edition*. San Francisco, CA:Benjamin Cummings.

It trickles back every year. The virus is here to stay and we're just doing everything we can to deal with it.

Cody Alcott

Viruses

Extracellular

Intracellular

Characteristics

Structure

Shapes

Replication

Lytic Cycle

Lysogenic Cycle

Viral Diseases

> **Measles**
> **Mumps**
> **Chickenpox**
> **Shingles**
> **Influenza**
> **Common cold**

Viruses' Role in Cancer

Cultivation of Viruses

Prions

Prion Diseases

> **Kuru**
> **Bovine Spongiform Encephalopathy (BSE)**
> **Creutzfeldt-Jakob Disease (CJD)**
> **Chronic Wasting Disease (CWD)**

Possible Prion Diseases

> **Alzheimer's**
> **Parkinson's**
> **Lou Gehrig's (ALS)**

Chapter # 2 – Chemistry of Microbiology

Bauman, Robert. (2009). *Microbiology With Diseases by Taxonomy, 3rd Edition*. San Francisco, CA:Benjamin Cummings.

Life exists in the Universe only because the carbon atom possesses certain exceptional qualities.

James Jeans - British scientist

Basic Chemistry

    Organic Chemistry

    Biochemistry

    Chemical Moieties

      Atoms

        Nucleus - protons + neutrons

        Electron shells or configuration

          1st shell = 2 electrons
          2nd shell = 8 electrons
          3rd shell = 8 electrons
          4th, 5th, 6th = 18 electrons each

        Outermost shell tendency

        Chemical reactions

        Valence

Elements

Isotopes

Molecules

Compounds

Chemical Bonds

Ionic bonds

Covalent bonds

Hydrogen bonds - weakest of all 3 bonds

Chemical Reactions

Collision theory

Activation energy
Water in Biological Systems

Polar molecule

Hydrogen bonding

Acids, Bases, and Salts

Acids

Bases

pH

Salts

Buffers

Biochemistry - 4 Major Groups

Carbon

Carbohydrates - Carbon, Hydrogen, & Oxygen

      Monosaccharides, Disaccharides, Polysaccharides

      Glycosidic bond

      Dehydration synthesis vs. hydrolysis reaction

      Examples and Functions

Lipids - Carbon, Hydrogen, & Oxygen

      Non-Polar Molecules - Hydrophobic

      Simple vs. Complex

      Waxes

      Steroids

Proteins - Carbon, Hydrogen, Oxygen, Nitrogen, & Sometimes Sulfur

Amino Acids

Primary Structure

Secondary Structure

Tertiary Structure

Quaternary Structure

Nucleic acids - DNA/RNA

DNA

Bases

RNA

Bases

Types

ATP

Chapter #5-Microbial Metabolism

Bauman, Robert. (2009). *Microbiology With Diseases by Taxonomy, 3rd Edition*. San Francisco, CA:Benjamin Cummings.

Discovery consists of seeing what everybody has seen, and thinking what nobody has thought.

Albert Szent-Györgi

Metabolism (Catabolism + Anabolism)

Catabolism

Anabolism

Enzymes

Chemical Reactions

Enzymes and Activation Energy

Enzyme Specific

Enzyme Components

Michaelis-Menten Equation $E + S = ES = E + P$

Enzyme Inhibition

Catabolism

ATP

REDOX Reactions - OIL RIG

Oxidation (atom, ion, or molecule loses electrons)

Reduction (atom, ion, or molecule gains electrons)

Glycolysis (Embden-Meyerhof Pathway)

Oxidation of Glucose

Citric Acid Cycle (TCA or Kreb's cycle)

Electron Transport Chain

Complete catabolic process produces net 38 ATP in prokaryotic cells & 36 ATP in eukaryotic cells

Fermentation - Anaerobic

Anabolism

Photosynthesis

Chemosynthesis

Chapter #6 Microbial Nutrition & Growth

Bauman, Robert. (2009). *Microbiology With Diseases by Taxonomy, 3rd Edition*. San Francisco, CA:Benjamin Cummings.

It is through science that we prove, but through intuition that we discover.

Henri Poincare'

Classification of Organisms Based on Nutrition

Energy Source

Carbon Source

Energy + Carbon Source

Oxygen Requirements

Obligate Aerobes

Obligate Anaerobes

Facultative Anaerobes

Aerotolerant Anaerobes

Microaerophiles

Physical Requirements

    Temperature

    pH

    Moisture

    Osmotic Pressure

Culturing Microorganisms

    Clinical Sampling

    Culturing Media

        Chemically defined

        Complex

        Enriched

        Selective

        Differential

        Anaerobic

Low or No Oxygen Requirements

Growth of Microbes (Numbers)

> **Binary Fission**

> **Generation Time**

> **Population Growth Curves**

Measuring Microbial Growth

> **Dilutions**

> **Direct Methods**

>> **Microscopic**

>> **Electronic counters**

>> **Plate count**

>> **Membrane filtration**

>> **Most probable number**

> **Indirect Methods**

>> **Active**

Chapter #7 Microbial Genetics

Bauman, Robert. (2009). *Microbiology With Diseases by Taxonomy, 3rd Edition*. San Francisco, CA:Benjamin Cummings.

We share 51% of our genes with yeast and 98% with chimpanzees-- it is not genetics that make us human.

Tom Shakespeare

Genetics

Genotype

Phenotype

Chromosomes

Plasmids

Flow of Genetic Information – Central Dogma of Molecular Biology

DNA Replication

Protein Synthesis

Transcription

Translation

Mutation

Types of Mutations

Spontaneous Mutations

Induced Mutations – Mutagens

Frequency of Mutation

Acquisition of New Genetic Information by Bacteria

Vertical vs. Horizontal

Transformation

Transduction

Conjugation

Chapter #9 Controlling Growth In-Vitro

Bauman, Robert. (2009). *Microbiology With Diseases by Taxonomy, 3rd Edition.* San Francisco, CA:Benjamin Cummings.

One sometimes finds what one is not looking for.

Alexander Fleming

Terminology of Microbial Control

    Sterilization

    Disinfection

        Disinfectant

        Antiseptic

        Pasteurization

    Degerming-Primarily Mechanical

    Sanitization-Safe Levels

    Suffix -stasis or-static

    Suffix -cide or-cidal

Microbial Death Rates

Action of Antimicrobial Methods and Agents

Physical Antimicrobial Procedures

    Heat related methods

    Dessication

    Filtration

    Radiation

    Osmotic pressure

    Refrigeration & freezing

Chemical Antimicrobial Procedures

    Phenol & phenolics

    Alcohols

    Halogens

    Oxidizing agents

    Surfactants

    Heavy metals

    Aldehydes

Chapter #10 Controlling Growth In-Vivo

Bauman, Robert. (2009). *Microbiology With Diseases by Taxonomy, 3rd Edition*. San Francisco, CA:Benjamin Cummings.

I have been trying to point out that in our lives chance may have an astonishing influence and, if I may offer advice to the young laboratory worker, it would be this – never neglect an extraordinary appearance or happening.

Alexander Fleming

History of Antimicrobials

Mechanisms of Antimicrobial Action

Selective Toxicity

Inhibition of Cell Wall Synthesis-M*ost Common*

Inhibition of Protein Synthesis-N*ext Most Common*

Injury to Plasma Membrane

Inhibition of Metabolic Pathways

Inhibition of Nucleic Acid Synthesis

Analogs Binding to Either Ligands or Receptors – Viruses

Clinical Considerations in Prescribing Antimicrobials

Definition of Ideal Antimicrobial Agent

Spectrum of Action

Efficacy

Static tests

Cidal tests

Bacteria & Drug Resistance

Mechanisms of Resistance

Methicillin Resistant *Staphylococcus aureus* – MRSA

Multiple Drug Resistant Organisms – MDRO - $\geq$ 3 antimicrobial agents

Retarding Drug Resistance

Antimicrobial

Chapter #14 Infection, Infectious Diseases, & Epidemiology

Bauman, Robert. (2009). *Microbiology With Diseases by Taxonomy, 3rd Edition.* San Francisco, CA:Benjamin Cummings.

Common things happen commonly, uncommon things don't, therefore if you hear hoofbeats assume horses, not zebras, but it is the occasional zebras which concern the microbe hunters.

A Medical Adage

Symbiotic relationship between different organisms

Mutualism

Commensalism

Parasitism

Microbial Antagonism

Indigenous Flora (normal)

Pathogens

Opportunistic Pathogens

Exposure & Invasion

Contamination & Infection

Portals of Entry

Adhesion of Microbes

Portals of Exit

Nature of Infectious Diseases

Manifestation of Disease

Etiology

Measurement of Pathogenicity

LD_{50}

ID_{50}

Virulence Factors

Extra cellular enzymes

Toxins-220 known bacteria toxins

Stages of Infectious Disease

Sources of Infectious Disease

Transmission of Infectious Disease

Contact

Vehicle

Vector

Classification of Infectious Disease

Epidemiology

Chapter #15 Innate Immunity – Defend Yourself

Bauman, Robert. (2009). *Microbiology With Diseases by Taxonomy, 3rd Edition*. San Francisco, CA:Benjamin Cummings.

Man knows in the end that he is alone in the indifferent immensity of the universe from which he has emerged by chance. Neither his destiny nor his duty is written down anywhere.

Jacques Monod

Overview of Immune System

A. *Non-Specific:* **Mechanical/anatomical barriers**
B. *Non-Specific:* **Chemical/physiological barriers**
C. *Specific:* **Adaptive barriers**

First Line of Defense: Non-specific

Mechanical/Anatomical Factors

Chemical/Physiological Factors

Second Line of Defense: Non-specific

Phagocytosis

Leukocytes

Mechanisms of Phagocytosis

Chemical/Physiological Factors

Inflammation

    Initiated

    Signs & Symptoms

    Function

    Fever

Chapter #16 Specific Defense – Adaptive Immunity

Bauman, Robert. (2009). *Microbiology With Diseases by Taxonomy, 3rd Edition.* San Francisco, CA:Benjamin Cummings.

I hope that some day the practice of producing cowpox in human beings will spread over the worl---when that day comes, there will be no more smallpox.

Edward Jenner

Acquired Immunity

Naturally Acquired Active Immunity

Naturally Acquired Passive Immunity

Artificially Acquired Active Immunity

Artificially Acquired Passive Immunity

Elements of Adaptive Immunity

Antigens

Cells, Tissues, & Organs of the Lymphatic System

Lymphatic vessels

Lymph nodes

Nose associated lymphatic tissue-NALT

Mucosal associated lymphatic tissue-MALT

B Lymphocytes

B cells-umoral response

Antibodies (immunoglobulins)

T Lymphoocytes

T cells-cell-mediated response

Helper T cells

Cytotoxic T cells

Humoral Immune Response

Cell Mediated Immune Response

Chapter #17 Immunization & Immune Testing

Bauman, Robert. (2009). *Microbiology With Diseases by Taxonomy, 3rd Edition.* San Francisco, CA:Benjamin Cummings.

The reward for work well done is the opportunity to do more.

Jonas Salk

History of Immunization

> **Variolation**

> **Vaccination**
> **Edward Jenner**

> **Louis Pasteur**

Active Immunization

> **Attenuated (Live) Vaccines**

> **Inactivated (Killed) Vaccines**

> **Sub-unit Vaccines**

> **Toxoid Vaccines**

Passive Immunization

Immune Testing

BIBLIOGRAPHY

Alcamo, E., & Elson, M. (1996). *The Microbiology Coloring Book*. San Francisco, CA: Benjamin Cummings.

API Testing – Biomerieux. (n.d.). Retrieved March 19, 2010, from http://www.biomerieux-diagnostics.com/servlet/srt/bio/clinical-diagnostics/dynPage?doc=CNL_PRD_CPL_G_PRD_CLN_11

Are You Cavity Prone? BioKit. (1999). Burlington, NC: Carolina Biological Supply Company.

Baron, S. (1996). *Medical Microbiology, 4th ed.* Retrieved March 30, 2010 from http://www.ncbi.nlm.nih.gov/bookshelf/br.fcgi?book=mmed&part=A512.

Bauman Robert. (2009). *Microbiology With Diseases by Taxonomy, 3rd Edition*. San Francisco, CA: Benjamin Cummings.

Bauman, Robert. (2007). *Microbiology with Diseases by Taxonomy, 2nd edition*. San Francisco, CA: Benjamin Cummings.

Bergquist, L., & Pogosian, B. (2000). *Microbiology Principles and Health Science Applications*. Philadelphia, PA: W.B. Saunders.

Brunvand, J.H. (1994). *The Big Book of Urban Legends*. New York, NY: Paradox Press.

Burton, G., & Engelkirk, P. (2000). *Microbiology for the Health Sciences, 6th edition*. Philadelphia, PA: Lippincott Williams and Wilkins.

Cappuccino, J., & Sherman, N. (2002). *Microbiology, A Laboratory Manual, 6th edition*. San Francisco, CA: Benjamin Cummings.

Centers for Disease Control and Prevention. (1995). *Unexplained Severe Illness Possibly Associated with Consumption of Kombucha Tea, Iowa 1995*. Morbidity and Mortality Weekly Report, 44(48), 892-893. Retrieved March 12, 2010 from www.cdc.gov/**mmwr**/preview/**mmwr**html/00039742.htm.

Centers for Disease Control and Prevention. (2009). *Health, United States, 2009*. Retrieved June 20, 2010 from www.cdc.gov/nchs/data/hus/hus09.pdf.

DeKruif, P. (1926). *Microbe Hunters*. New York, NY: Harcourt Brace and Co.

Fiedel, D. (2002). *Fiedel's Official Ghost Guide to Lancaster County, Pennsylvania*. Lancaster, PA: Lancaster Historical Society.

Girstenblith, Meg Rebecca (Winter 2000). "A City Besieged: The 1918 Pandemic in Lancaster, PA". *Lancaster County Historical Society*. 102(4). P. 138.

Glo Germ Products. (n.d.). Retrieved March 15, 2010, from http://www.glogerm.com/.

Greenaway, K. (1881). *Mother Goose or the Old Nursery Rhymes*. London, UK: George Rutledge & Sons.

Hach Company. (2000). *Methods 8241 & 8242: For Water and Wastewater*. Retrieved July 12, 2010, from http://www.water-research.net/Waterlibrary/watermanual/platcount.pdf.

Hairston, R.C. (1999). *Biology 221-Student Study Guide & Laboratory Manual*. Dubuque, IA: Kendall/Hunt.

Johnson, T., & Case, C. (2004). *Laboratory Experiments in Microbiology, 7th edition*. San Francisco, CA: Benjamin Cummings.

Katz, S. & Leyva (2008). *Modifying the Kirby-Bauer Antibiotic Susceptibility Exercise to Promote Active Learning*. ASMCUE . May 30, 2008. Endicott College. Boston. MA.

Leboffe, M., & Pierce, B. (2002). *Microbiology Laboratory Theory and Application*. Englewood, CO: Morton Co.

Leland, Anne. American War & Military Operations: Lists & Casualties. Retrieved March 23, 2010, from www.fas.org/sgp/crs/natsec/RL32492.pdf.

National Yogurt Association (NYA). Retrieved March 30, 2010, from http://aboutyogurt.com/index.asp?bid=5.

Nguyen, Q. (2009) *Hospital-Acquired Infections*.Retrieved March 30, 2010, from http://emedicine.medscape.com/article/967022-overview.

Pediatric Dental Health. (2005). Retrieved March 30, 2010 from http://www.dentalresource.org/topic56scarletfever.html.

Rapoza, M., & Kreuzer, H. (2004). *Transformations: A Teacher's Manual*. Burlington, NC: Carolina Biological Supply Company.

Redbone Heritage Foundation. Retrieved March 23, 2010, from http://.redboneheritagefoundation.com/Chronicles/Disease%20Epidemics%20in%20Early%20America.ht

Root Beer. (n.d.). Retrieved August 13, 2010, from http://www.root-beer.org/.

Shank's Extracts Inc. Flavoring Extracts and Syrups. 350 Richardson Drive, Lancaster, PA, 17603-4034. 717-393-4441.

Staley, J., Gunsalus, R., Lory, S., Perry, J. (2007). *Microbial Life, 2nd Edition*. Sunderland, MA: Sauter Associates.

Tierno, P. (2001). *The Secret Life of Germs*. New York, NY: Pocket Books.

Tortora, G., Funke, B., & Case, C. (2002). *Microbiology: An Introduction, 7th edition*. San Francisco, CA: Benjamin Cummings.

Transformations: a Teacher's Manual. (2004). Burlington, NC: Carolina Biological Supply Company.

Varasdi, A. (1989). *Myth Information*. New York, NY: Ballantine Books.

Warrick, J. (2010). "FBI Investigation of 2001 Anthrax Attacks Concluded; US Releases Details". Washington Post. http://www.washingtonpost.com/wp-dyn/content/article/2010/02/19/AR2010021902369.html.